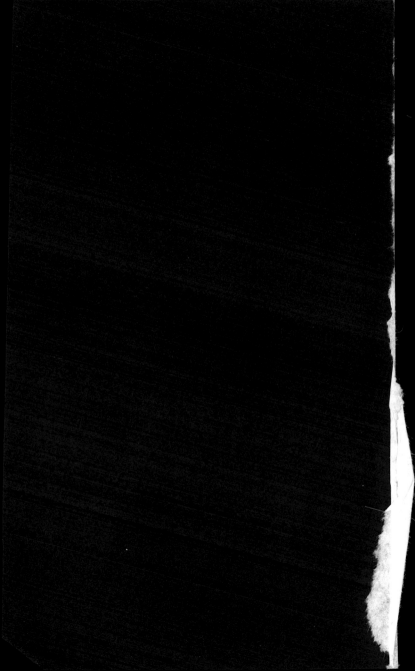

ALEXANDER NIMMO'S INVERNESS SURVEY & JOURNAL, 1806

Alexander Nimmo's Inverness Survey & Journal, 1806

~ Edited by Noël P. Wilkins ~

RIA

Alexander Nimmo's Inverness survey and journal, 1806

First published 2011

by Royal Irish Academy
19 Dawson Street
Dublin 2

www.ria.ie

The authors and publisher are grateful to the National Library of Scotland, the Royal Dublin Society and the Highland Folk Museum for permission to reproduce material in this book.

ISBN 978-1-904890-74-4

British Library Cataloguing in Publication Data. A CIP catalogue record for this book is available from the British Library.

Printed in the UK by MPG Books Ltd.

CONTENTS

Acknowledgements

Many persons contributed to the preparation of this volume. Mr Martyn Wade, National Librarian of Scotland, kindly gave access to, and permission to transcribe, the original MSS which are in the Library's care. Ms Sheila Mackenzie, Senior Manuscript Curator, facilitated and encouraged the project from the beginning and this is very much appreciated, as is the unfailing courtesy and assistance of the Library staff in the George IV Bridge reading room. Likewise Dr Peter Heffernan, MRIA and CEO of the Marine Institute of Ireland, actively encouraged and supported the endeavour from the outset and Professor Nicholas Canny, PRIA was also a constant support. Professors C.W.J. Withers, FRSE, University of Edinburgh, gave invaluable advice at the preliminary and final stages.

Ms Grace Uhr transcribed the first draft into a computer file from a photocopy of the original with consummate skill, care and accuracy, thereby making later drafts much easier to prepare and compare with the original MSS.

Ms Helena King, publications officer of the RIA, helped with the progress of the project from its initiation through the preparation and sponsorship processes to final completion.

We wish to thank the following for their assistance over a number of years in accessing material and sources in their care: The National Library of Scotland; the National Archive of Scotland, Edinburgh; the National Archives of Ireland, Dublin; Highland Council Archive, Inverness; the Highland Council Library Service and the Inverness Royal Academy Archive.

Special thanks go to the Trustees of the National Library of Scotland for permission to publish the transcript and figures taken from it. The Royal Dublin Society kindly gave permission to reproduce the photograph of the bust of Nimmo in the Society's possession. It is the only known image of him.

The publication would not have been possible without generous sponsorship from the Marine Institute of Ireland, the Royal Society of Edinburgh and the Royal Irish Academy for which we are truly grateful.

The National University of Ireland, Galway, provided continuing support without which the project could never have advanced.

FOREWORD

It is with pleasure that we introduce *Alexander Nimmo's Inverness survey and journal, 1806* to a wider public. When the editor Noël P. Wilkins suggested that the Royal Irish Academy and the Royal Society of Edinburgh come together to publish, for the first time, the account of our shared Member's 'perambulations' in Inverness-shire, it made an attractive proposal.

Both the RIA and the RSE originated for similar purposes and under similar scientific circumstances. After the RSE was founded in 1783 for the 'Advancement of Learning and Useful Knowledge' in Scotland, the benefits of forming an Irish analogue became clear. Just two years later in 1785, the RIA was established under similar rubric. The tradition of support and exchange of ideas has always animated the relationship between the two bodies. Alexander Nimmo put it well himself in his opening of a paper published in the *Transactions of the Royal Irish Academy* on 27 October 1823, titled 'On the Application of the Science of Geology to the purposes of Practical Navigation':

> It is an old and perhaps a trite observation, that all the various branches of science are calculated to throw light upon each other; and hence, that an extended acquaintance with the several departments of natural knowledge is the surest way to attain to eminence in the pursuit of any one. In a society like this, composed of persons of different pursuits, and whose ideas are directed to a variety of objects, it is perhaps one of the greatest advantages, that, in any new enquiry, we may be enabled to draw from stores of information that are beyond the reach of any individual mind.

Nimmo's talents did not lie in one narrow discipline. He began his career as a schoolmaster before turning to the practical skills of civil engineering, which he by no means limited himself to, fascinated as he was in the different fields and aspects of rural life explored in this volume. Given his versatility and broad vision, he foresaw the possibilities of the learned society as a forum in which to exchange and generate knowledge and ideas. The RIA and the RSE continue this tradition, with a grant scheme aimed at fostering collaboration between researchers in both Ireland and Scotland.

Several Members have been elected to both institutions, covering both the arts and sciences. These include William Bald, the Scottish land surveyor and contemporary of Nimmo, whose principal achievement was in Ireland and, in the more recent past, the distinguished paleobotanist Sir Alwyn William and the eminent chemist, Charles Kemball, who both served as Presidents of the RSE. Today John Brewer, a sociologist with expertise in peace processes, and Tom Devine, a leading Scottish historian are also members of both.

Alexander Nimmo's legacy differs in each country. In Ireland he is chiefly remembered for his engineering innovations in Connemara and Kerry under the Famine Relief Act of 1822, creating safe havens for boats by building piers, founding the village of Roundstone and building road infrastructure, including the stunning carriage road from Maam Cross to Leenane. In Scotland he is remembered as Rector of Inverness Academy and, crucially, for his public work surveying and mapping the boundaries of Inverness-shire. He undertook the commission on a vast scale in 1806, when only in his early twenties.

200 years after his election as a Fellow of the RSE, Nimmo's work remains stimulating and relevant, allowing scholars and engineers to draw from the 'stores of information' he recorded as they go about planning and reshaping the infrastructure of Scotland and Ireland.

We look forward to continued active and productive collaboration between the RIA and the RSE into the future.

Nicholas Canny, PRIA
Sir John Arbuthnott, MRIA, FRSE, President-elect RSE

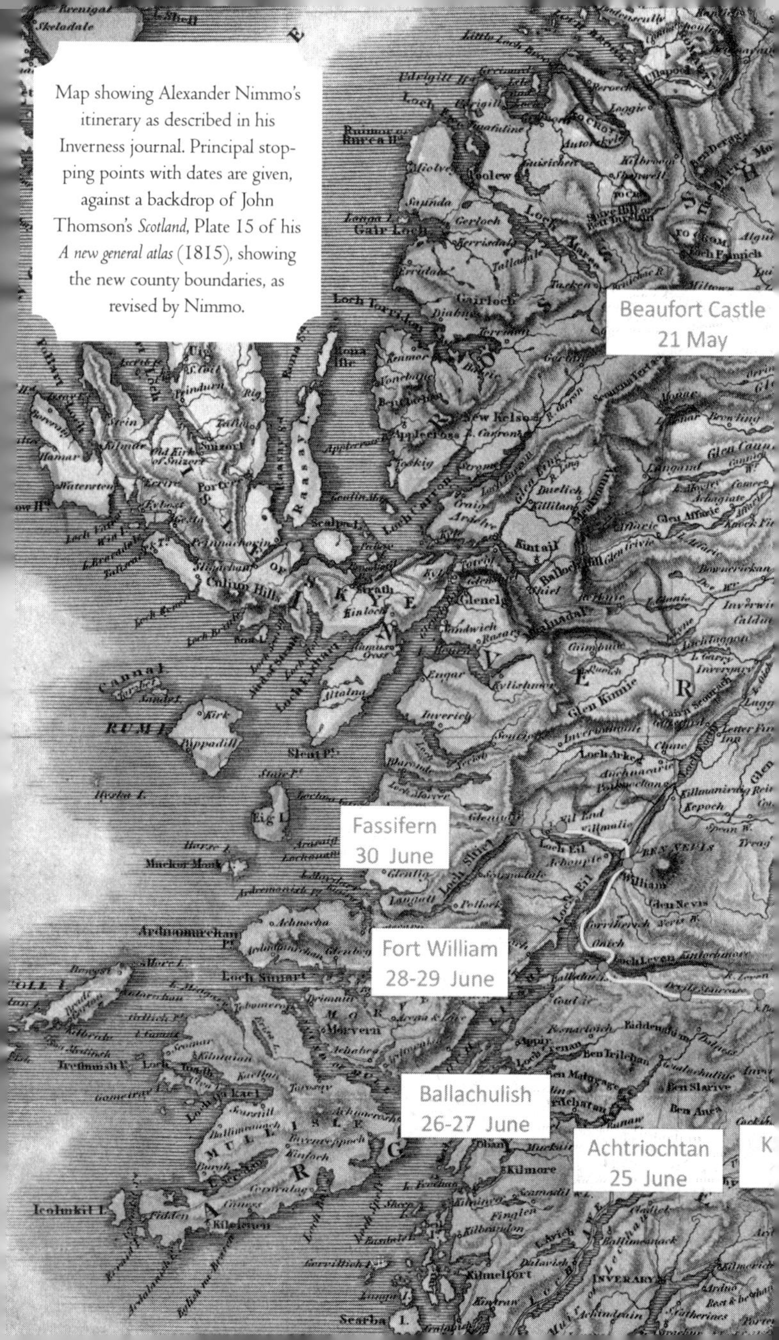

Map showing Alexander Nimmo's itinerary as described in his Inverness journal. Principal stopping points with dates are given, against a backdrop of John Thomson's *Scotland*, Plate 15 of his *A new general atlas* (1815), showing the new county boundaries, as revised by Nimmo.

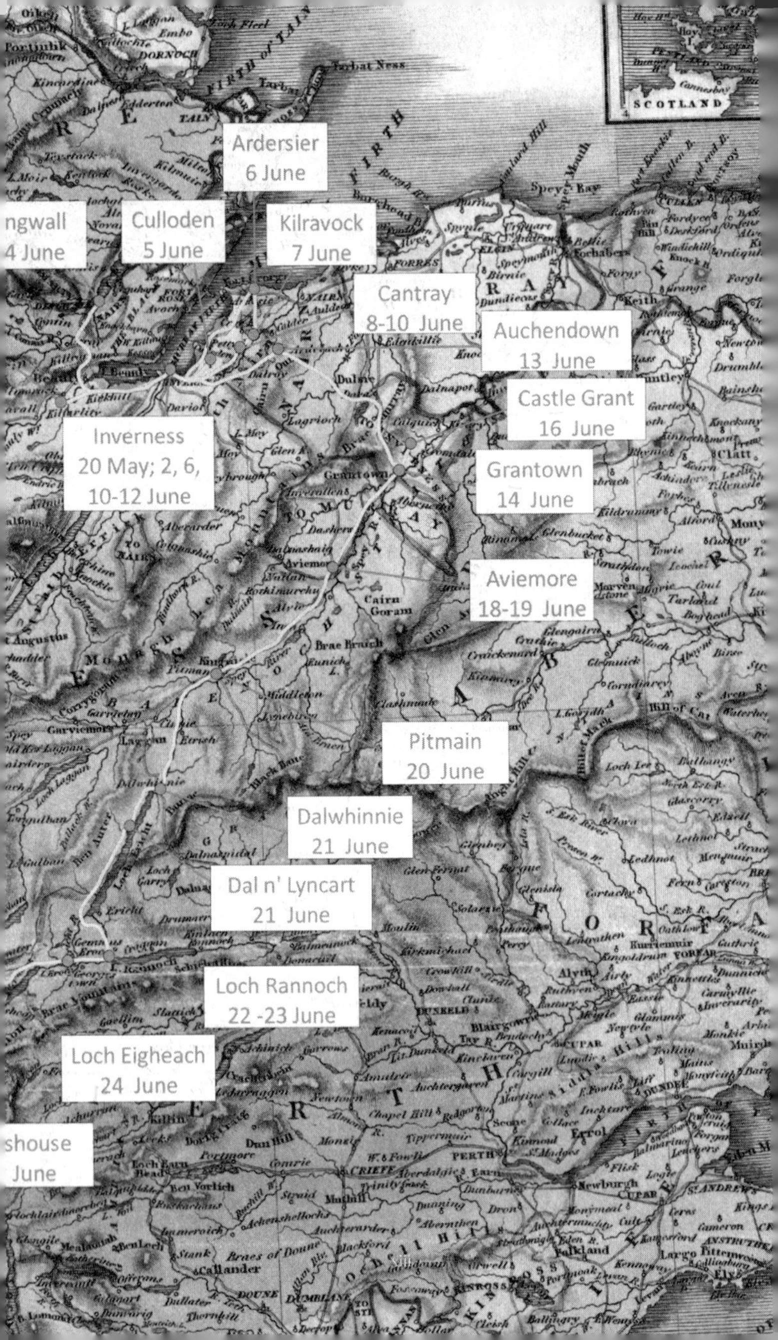

SCOTLAND
Ardersier
6 June
ngwall
4 June
Culloden
5 June
Kilravock
7 June
Cantray
8-10 June
Auchendown
13 June
Castle Grant
16 June
Inverness
20 May; 2, 6,
10-12 June
Grantown
14 June
Aviemore
18-19 June
Pitmain
20 June
Dalwhinnie
21 June
Dal n' Lyncart
21 June
Loch Rannoch
22 -23 June
Loch Eigheach
24 June
shouse
June

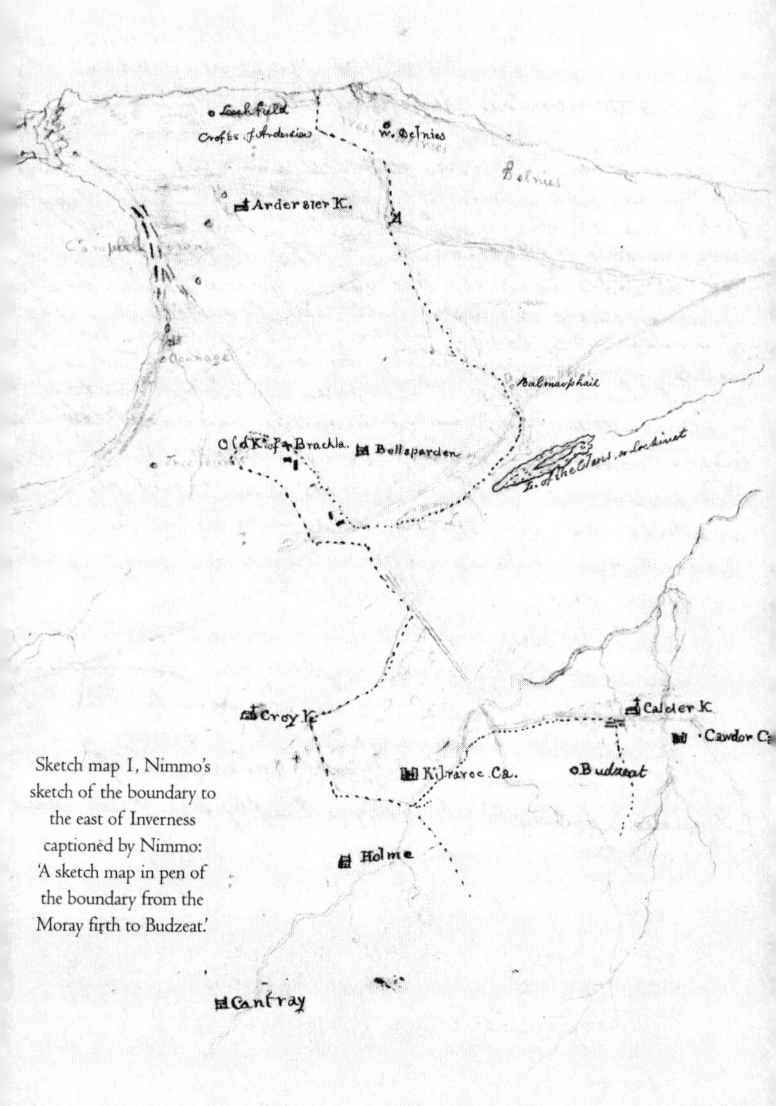

Sketch map I, Nimmo's sketch of the boundary to the east of Inverness captioned by Nimmo: 'A sketch map in pen of the boundary from the Moray firth to Budzeat.'

The Heads of The Beaulie River &c.
with the County Boundary in Glasletir

Strath Monar
L. Monar
Strath Farrar
R. Farrar
Glen Cannich
R. Cannich
Muchrach
L. Molardich
L. Benevian
L. Affarie
Glen Grivie
Glen Elchaig
L. Achaghuirn
R. Carron
Struy
Comar
falls
4 Miles p. Inch
Glen Sheal

Sketch map 2, four miles to one inch. Captioned by Nimmo: 'The Heads of the Beaulie River etc. with the county boundary in Glasletir.'

Chapter I ~ Introduction

The original manuscript and its transcription[1]

~ Noël P. Wilkins ~

Introduction

In the late eighteenth century social and economic life in the Scottish Highlands underwent a radical transformation.[2] In brief, the old social order—a clan-based, quasi-feudal system—was passing away. While the landowning lairds became increasingly anglicised and gentrified, their tacksmen first, followed later by their lower tenants, were forced to emigrate.[3] Emigration freed up land that the lairds could then rent out (at greatly inflated prices) to willing lowlanders and others, increasing the lairds' monetary wealth even as they displaced their erstwhile loyal clansmen. The lairds and the new leasees stocked the hills and rough pastures with sheep, further squeezing the few poor tenants who remained. Thomas Telford summed up the transformation pithily: '...the lairds have transferred their affections from the people to flocks of sheep and the people have lost their veneration for the lairds'.[4]

Writers, politicians and academics, even some military men, had been describing the predicament of the Highlands for years: individual glens were isolated; road communication was extremely poor or non-existent; there was no paid work and few tools or skills; and the inhabitants—'...[who] may be considered as prisoners strongly guarded by impassable mountains on one side, by swamps and furious torrents on the other...'[5]—were sinking into increased isolation, poverty and despair. In 1786 the British Fisheries Society was founded with a view to encouraging the fishing industry by the construction of new fishing villages, and later by championing the case for building new roads. Telford was appointed engineer to the Society and his visits to the Highlands informed his professional opinion of what was needed to relieve the situation. Still, little positive action was taken by the government until, at the close of the century, it finally became alarmed at the likely military repercussion of the large-scale emigration of experienced and tested Highland fighting men. That military considerations were important in the management of the Highlands is attested to in Robert Preece's essay in this volume, where he points out that topics such as fortifications, military architecture and practical gunnery were embedded in the curricula of the academies at the time—even though the pupils were only six to 14 years of age![6]

Telford was commissioned by government to survey the Highlands and islands and to devise improvements to internal communications and new coastal facilities. He made two surveys in 1801 and 1802, in the second of which he commented on the changes that sheep farming had caused and how this had influenced emigration: '...[sheep farming] not only requires much fewer people to manage the same tract of country, but in general an entirely new people who have been accustomed to this mode of life are brought from the southern parts of Scotland.'[7] As for the native Highlanders '...about three thousand persons went away in the course of the last year and if I am rightly informed three times that number are preparing to leave the country in the present year.'[8] The official response to these alarming reports was rapid and by July 1803 two separate commissions—the Parliamentary Commission for making Roads and building Bridges in the Highlands of Scotland and the Commission for the Caledonian Canal—had been set up. Telford was appointed engineer to both, and John Rickman was appointed secretary.

What these commissions did was to give coherence to schemes for improvements in land and water communication throughout the north while providing paid employment in public works to otherwise impoverished highlanders, under the direction of the most eminent civil engineer of the time. The Caledonian Canal would be a benefit to the whole nation, providing an inland communication from the eastern to the western ocean for trading and naval vessels, thereby avoiding the treacherous passage through the Pentland Firth. As such, its total cost rightly fell to be borne by the national exchequer. On the other hand, Telford's proposed schemes for harbours, roads and bridges would benefit in a measurable way the affected counties and local proprietors and in this circumstance the beneficiaries were to be levied a portion of the total cost. This decision necessitated an accurate delineation of the county boundaries (which were often part of the boundaries of individual estates), so that the levies could be applied equitably. Telford claimed in 1819 that there had been made '...about 1,000 miles of road and about 1,500 bridges beside a very considerable number of harbours, landing-piers and ferry-piers' in the Highlands and western part of Scotland, so the anticipated costs were not insignificant.[9]

In 1806, on Telford's recommendation, Alexander Nimmo, then Rector of Inverness Academy, was commissioned to undertake the necessary survey of the boundaries of Inverness-shire and to insert them on a new draft map of Scotland. At the time, the shires were not always single, integrated entities but sometimes consisted of a number of separate 'insulated' districts not contiguous with one another. Cromarty, for instance, comprised about 12 insulated districts scattered in the County of Ross. In true scholarly fashion, Nimmo first researched and compiled a historical statement, drawn from ancient and arcane sources, of the origins of the Highland counties that the Commissioners published in 1807 in their third annual report.[10]

In the course of his survey Nimmo maintained a personal journal in which he recorded a variety of observations and data. In his own words, it was not drawn up with any methodical or connected arrangement. It is, nevertheless, a valuable and interesting first-hand contemporary account of many aspects of life and change in Inverness-shire and the Highlands, at a most important time in their history. It is also a seminal document in Nimmo's personal transformation from schoolmaster to engineer. His experience of the survey would lead ultimately to his resignation from the Rectorship in 1811 to join the Commission for the Bogs of Ireland as an engineer. His journal therefore gives us an insight into the genesis of the approach that he would bring to his surveys of County Kerry and Connemara, Co. Galway in 1811–12 and 1813, respectively, and later to all of the west of Ireland where he was government engineer from 1823 to 1832. In his Irish engineering career, which would last until his death in 1832, Nimmo appears to have emulated Telford's iconic Scottish role: what Telford was to the Highlands of Scotland, Nimmo would strive to be to Ireland.

On completion of the Inverness survey he sent his journal, some detailed sketches and a map of the Black Isle to John Rickman, secretary to the Commissioners, together with the outline map he had earlier received, now with the confirmed boundaries delineated and the adjacent counties coloured in. Rickman forwarded them to Aaron Arrowsmith, who was to engrave the new map of Scotland for the Commissioners. The journal with its three rough sketches and the accompanying letter to Rickman are all that appear to have survived of this material. Now is the first time that these manuscripts, which are conserved in the National Library of Scotland, have been published.

The accompanying essays seek to place the man and his survey in the context of his career as a schoolmaster, the historical events of the time in Scotland and Ireland,

and his subsequent, successful engineering career in Ireland and Britain.

Collation

The complete manuscript is bound in cream-coloured cards, 260mm tall by 205 mm wide and comprises three separate manuscript items written on 56 folios, numbered consecutively in a modern hand, and titled as follows:

- 'Journal along the North East and South of Inverness Shire. Ends at Fort William'. Signed by Alexander Nimmo;
- 'Scotch Boundary's [*sic*]'. A letter to John Rickman, signed by Alexander Nimmo, dated 12 October 1806;
- 'The Heads of the Beaulie river etc with the County boundary in Glasletir'. A sketch map in pen and ink with accompanying text, initialled A.N. and dated 14 August.

The journal

The journal comprises 50 folios, 230 mm tall by 189 mm wide, numbered 1 to 50. The narrative is written entirely on the rectos; the versos of most folios are blank, except where ancillary information—glosses on Gaelic etymology and/or astronomical observations—is recorded.

The folios of the journal appear to comprise two gathers or 'copybooks':

- Folios 1–26. A copybook comprising a single gather of 13 sheets folded once to give 26 folios. Folios 1 to 7 and 20 to 26 are watermarked but without a maker's name. Folios 8 to 13 and 14 to 19 are watermarked 'W. Kingsford'. This is the first 'copybook'.

- Folios 27–50. A copybook comprising a single gather of 12 sheets folded once to give 24 folios. Folios 27 to 32 and 45 to 50 are watermarked 'John Fellows 1805'. Folios 33 to 44 have no clear watermark. This is the second 'copybook'.

Scotch Boundary's [sic]

This consists of a single gather of 3 folded sheets forming 6 folios, 250 mm tall by 199 mm wide, numbered 51 to 54 inclusive. The sheets are watermarked 'Buttenslaw [?] 1801'. Folio 51 is a title page with the words 'Scotch Boundary's [*sic*]' inscribed. Beneath this, in a different hand, the following entries are written:

World	on roller	1–16–0
North America	do	1–10–0
Aus [?]	do	1–5–0
		4–11–0

Neither the title nor the entries on this folio appear to be in Nimmo's hand and the misuse of 'Boundary's' in place of 'Boundaries' is unexpected from a schoolmaster. But folio 51 is certainly part of the same sheet as folio 54, which bears Nimmo's signature (see below).

Someone else may have written the misspelt title on the blank folio to identify the document; the map cost entries are entirely unrelated to the title and to the letter itself. Alternatively, soon after his arrival in the Inverness Academy, Nimmo had requested and been authorised to purchase globes. Later Matthew Adam, Nimmo's successor as Rector, was granted authority to spend money on globes and new maps, including 'roller maps of Europe, Asia, Africa, North America and South America' at a cost of £7–17–6. The entries on the title page may therefore be costings of some of these maps recorded by Nimmo in 1806, which he or another person may have written inadvertently on the title page before sending the letter to Rickman. Or maybe Rickman had sought out information for Nimmo and jotted it down on the letter. All this, however, is just speculation.

Folio 51 verso is blank.

Nimmo's letter starts on folio 52 recto and continues on the recto and verso of folios 52 and 53 and on the recto only of folio 54. Folio 54, with Nimmo's signature, is part of the same sheet as folio 51, so the title page is integral to the letter.

Folio 54 verso is blank.

Heads of the Beaulie River etc

This comprises a single sheet, folded once to give 2 folios, 257mm tall by 202 mm wide. The sheet is watermarked but without the maker's name or date.

Folio 55 recto contains the pen and ink sketch map and part of the accompanying text.

Folio 55 verso is blank.

Folio 56 recto continues the accompanying text and ends with a postscript initialled A.N. and dated 14 August.

Folio 56 verso has the name John Rickman inscribed on it and the whole of this item shows evidence of having been folded into sixths at one time, i.e. folded as a letter, with Rickman's name on the outside.

Because there is no evidence of such folding in the journal or the Scotch Boundary's [sic] manuscripts, and the items have different dates, the 'Beaulie River' document is not integral to either the letter or the journal, although all three items are incontrovertibly by Nimmo. The letter, on folios 51 to 54, would appear to be a covering letter that possibly accompanied the journal, folios 1 to 50. The Beaulie River document, folios 55 and 56, appears to be an additional communication to Rickman by Nimmo, possibly in response to a query.

Notes on the transcription

The text was transcribed and typed from a photocopy of the original manuscript. The clean typescript was then carefully proof-read and corrected by reference to the photocopy. Ambiguities and uncertainties were noted and marked on the typescript. These were then corrected by direct comparison with the original manuscripts in the National Library of Scotland. The typescript and original manuscripts were read together and every ambiguity or uncertainty was individually examined and corrected. There were very few ambiguities, since the original manuscript is generally clear and easy to read (see p10), and only two or three cases proved impossible to decipher.

Nimmo's punctuation and use of capital letters do not conform well to the conventions of today. He used capitals liberally for common names but did not always commence sentences with a capital. His spelling was similarly flexible. For example, he showed no consistency with names like 'Macpherson' and 'McPherson', 'Mackenzie' and 'McKenzie', nor with words and

Thursday.—
Meantime I went to Culloden & was shown
by Mr Kinloch factor for Forbes of Culloden some plans
of his farms on the water of Nairn - several of which
are situated in the county of Nairn ——

I received the telescope by the carrier from Aberdeen
whither it had come by the Mail — this I afterwards
found one of the most valuable instruments I could
have in my possession — The Sextant & Horizon
had already arrived by sea — from their bulk they
became very inconvenient in travelling — the trouble
of unlashing them when fixed on horse back — prevented me from taking several observations that
might have been of service in constructing or correcting the map — fortunately in marking the outline I seldom found them necessary — the number
and accuracy of the Physical positions on the map
enabled me in many cases to speak with more
precision than even many of my guides when
we were not on the spot ——
I took along with me a small pocket compass
it was not fitted with sights but was otherwise of
the greatest service — in so much that at length
in drawing a slight sketch of any little district
and in recognising objects when on elevated stations
I came at length to rely chiefly on that and the
Telescope ——

Example of Alexander Nimmo's writing from his journal
[Folio 11 recto, National Library of Scotland, Adv MS 34.4.20]

abbreviations, with variations such as 'although', 'altho' and 'though'. His spelling of many topographical names is understandably erratic, since he did not speak or write Gaelic (although he displayed a lively and accurate interest in the etymology of Gaelic words and phrases). For example, for Loch Erich he uses 'Eruch', 'Eroch' and 'Errich', often using two or more alternative spellings on the same page; mountain names starting with 'Sgur' are variously written 'Sgur', 'Scur', or 'Scor'. Even with clear English names like Kingshouse he uses many variants, for example 'Kingshouse', 'Kings house', 'Kings House' and 'King's house'.

This transcription retains Nimmo's spelling in all cases. Only the punctuation and use of capitals have been modified, and then only sparingly, to make the text more easily readable. Where words or names are uncertain they are followed by [?] or given by ------- with an explanation such as '[indecipherable]' or '[blank]'. Where common words are spelt differently from modern practice, or misspelt, they are followed by '[*sic*]', in order to distinguish these from unintended errors in transcription or mere typographical errors.

The numbers in brackets at the top of the text indicate the folio numbers written on the original manuscript folios. Rectos are indicated by numbers only, e.g. {f. 1}; when text occurs on the verso of any folio, this is indicated in the numbering e.g. {f. 6 verso}. Versos are placed in the sequence they hold in the original manuscript, even when this means that the continuation of the text of the rectos appears interrupted, e.g. {f. 6} is followed by {f. 6 verso}, followed by {f. 7} even when the text runs directly from {f. 6}to {f. 7}.

Explanatory notes are referenced by superscripts in the text and explained in full in the footnotes.

In the place and topographical features index, the page numbers given are the folio numbers in the original manuscripts, not those of the pages of this volume.

Notes to Chapter I

[1] Alexander Nimmo 'Journal along the North East and South of Inverness-shire. Ends at Fort William' (Inverness, 1806). Adv MS 34.4.20, National Library of Scotland. Manuscript reproduced in this volume.

[2] This is detailed by James Hunter in his essay in this volume, Chapter III

[3] A.R.B. Haldane, *New ways through the glens: Highland road, bridge and canal makers of the early nineteenth century* (Edinburgh, 1962).

[4] Thomas Telford to Andrew Little, 18 February 1803, quoted in A. Gibb, *The story of Telford* (London, 1935), 72.

[5] J. Knox, *A tour through the Highlands of Scotland and the Hebrides Isles in 1786* (1786, London, Edinburgh, Glasgow), quoted in Haldane, *New ways through the glens*, 12.

[6] See Robert Preece's essay in this volume, Chapter IV.

[7] T. Telford, 'Survey and report 1803', quoted in Haldane, *New ways through the glens*, 21.

[8] Thomas Telford, 'Evidence' in 'Second Report from the Select Committee on the State of Disease and Condition of the Labouring Poor in Ireland', British Parliamentary Papers (BPP) 1819 (314) Vol. VIII, microfiche (mf) 20.67, 33.

[9] Telford, 'Evidence'.

[10] Alexander Nimmo, 'Historical statement of the erection and boundaries of the shires of Inverness, Ross, Cromarty, Sutherland and Caithness', in 'Third Report of the Commission for the Highland Roads and Bridges', Appendix U, BPP 1807 (100) Vol. III, mf 8.12–3.

CHAPTER II

ALEXANDER NIMMO: RECTOR OF INVERNESS ROYAL ACADEMY

~ Robert Preece ~

Alexander Nimmo was the Fourth Rector of Inverness Royal Academy, taking up office in the autumn of 1805. The Academy had been founded in 1792 and paid for by private subscription, only 13 years before Nimmo, aged about 22, was to take over as Rector (a Scots word for headmaster).

Scottish schools and academies up to 1800

The opening of the Inverness Royal Academy was the start of the third phase of a long history of education in Inverness. From the thirteenth century up to its disbandment about the time of the Reformation in 1560, Inverness Academy had a Dominican Priory, with an associated 'song school'. Boys were taught Latin and singing so that they could provide music at the daily services in the church. This song school seems to have evolved into a grammar school, perhaps in the late sixteenth century, but the evidence for the transition is scanty. A grammar school, where the term 'grammar' meant Latin grammar, was generally the main school of a Scottish town. The instruction would be largely or entirely centred on the classics, with Latin, and perhaps Greek, featuring. As a result, it was only the brighter and better-off pupils who would attend, often from about age six or seven. Attending grammar school was normally a prerequisite to attending a university, generally from about age 14 or 15, and then moving into a professional career such as law, teaching or as a minister of religion. Towns might have had other schools teaching basic writing and reading, mathematics and sometimes other subjects. Most pupils were boys, but in some towns girls would also attend. Many of the landed families, however, would have had private tutors for educating their children, or else they would have been sent away for education to a school run along the English public school lines, perhaps in central Scotland.

In late-eighteenth-century Scotland the concept of academies, where one institution taught a range of subjects to young people, was a relatively new one. It was partly a response to the period of the Scottish Enlightenment (roughly covering the eighteenth and early nineteenth centuries), and partly a response to the need for young people to be trained in commercial subjects such as book-keeping, in line with Britain's increasing role as a nation trading with its colonies and later with its empire. Pupils attending an academy would vary in age from perhaps age six to age 14 or 15. They would not necessarily attend a course of study covering the full range of subjects, but would opt in to those subjects that their parents thought appropriate. Individual classes would also tend to have pupils of a range of different ages. The first Scottish schools to be named academies were in Ayr (in 1746) and Perth (in 1761), but they operated on rather different principles from later academies. It was not until 1786 that Dundee Academy was established on the lines that many other towns were to follow. Fortrose, in the Black Isle (to the north of Inverness), set up their rather small academy in 1791 with only three teachers. Nimmo's first teaching appointment was in 1802 as second master in that school.

In Inverness the town's grammar school continued in operation until the summer of 1792, when it closed to make way for the opening of the new Academy, the fifth to be established in Scotland. A new building had been erected in New Street, later to be called Academy Street, then on the edge of the town. The following year the Academy was to gain its 'Royal' title by purchasing a charter from King George III at a cost of almost £180 sterling. However, the 'Royal' part of the name was little used until towards the end of the nineteenth century.

When Inverness Academy was first opened, five teachers were employed, and the subjects taught were:

- English—'to be taught grammatically';
- Classics;
- Writing, arithmetic and book-keeping—writing to be taught to the scholars of the grammar school from noon to 1pm, and from 1pm to 2pm to 'young ladies who cannot attend at the ordinary school hours';

- Mathematics—Euclid, algebra, trigonometry, mensuration, geography with the use of globes, navigation with lunar observations, naval, civil and military architecture, practical gunnery, fortification, perspective and drawing; with the geography taught between 1pm and 2pm;
- Natural and civil history, natural philosophy, chemistry and astronomy.[1]

The teaching of (Scottish) Gaelic did feature when a teacher (and funds) was available. French and Greek also appeared as subjects from time to time. However, the key subjects were to change little in the early years, as is shown in an unsigned report, dated 1810, probably produced for the directors of Perth Academy, which compared the two schools.[2] It will be noted that girls attended Inverness Academy from the outset, although not in large numbers.

Through his work outside education, Nimmo was to show special practical and theoretical ability in many of the topics mentioned in the mathematics course, such as fortification and civil and military architecture. It must be remembered that this was a time of major warfare across continental Europe. Two of the local school staff were re-employed in the Academy when it first opened, one (for classics) from the grammar school, the other from the local school which had been teaching writing, arithmetic and book-keeping. The local teacher of English was also recommended as suitable, but did not get a job at the new Academy, as the directors wanted 'a native of England', apparently as they regarded a Scottish accent as unacceptable or inferior.

Alexander Nimmo, according to some early biographies, was born in Kirkcaldy, Fife, but was actually born in nearby Cupar in 1783.[3] His father was originally a watchmaker, but subsequently kept a hardware store in Kirkcaldy,[4] where Nimmo may have attended the burgh school, although this cannot be confirmed. In mid-1796, aged 13, he was awarded one of the Bayne bursaries, valid for four years, to attend the nearby University of St Andrews. These bursaries, whose patron was William Ferguson of Raith (on the western edge of Kirkcaldy), were worth £120 Scots (£10 sterling.) a year,[5] and were available for students whose families were fairly poor. The bursary was paid out by the magistrates of Cupar, suggesting that Nimmo's family still had links to that town. He spent two years at St Andrews, registering in early 1797, and another two years at Edinburgh University. Attendance at university from about age 14 was common at that time. Nimmo became an accomplished scholar in Latin and Greek as well as in various branches of mathematics and natural philosophy (physics). There is no evidence that he actually graduated from university, but few students in that era graduated formally.

Having been a tutor in Edinburgh, Nimmo became, in early 1802 when he was aged 18 or 19, second master at Fortrose Academy, teaching mathematics.[6] His salary was £35 sterling a year, plus the class fees from his 15 students; he also oversaw a boarding house for scholars, as was common at that time. In 1805 a new Rector brought a harsher regime to that Academy, but Nimmo seems to have been keen to move on even before this new appointment. Following the unexpected death, in early 1805, of Alexander Macgregor, the Third Rector of Inverness Academy, Nimmo applied for the job even before it was advertised.[7] The two men almost certainly knew each other, as they were both mathematicians with similar academic interests. There would have been few other similarly qualified people in the Highlands at that time. The mathematics curricula then taught at both Fortrose and Inverness Royal Academies seem very similar, reflecting the needs and interests of the era.

The Inverness Academy directors, despite Nimmo's application, decided to advertise in the Edinburgh newspapers before filling the vacancy. Four candidates responded: Nimmo, John Tulloch (already on the staff), a Mr Pollock from Glasgow and a Mr Adie from Nielston, near Glasgow. At their May meeting the directors decided to invite some of the professors at Edinburgh University to make the choice on their behalf. In some sources details of the appointment procedure are misleading, as information on the length of the examination and who the examiners were is presumably correctly given in the copy letter in the directors' minutes book after the meeting of 13 July 1805.[8] The letter reads:

Edinburgh, 27th June 1805

Report Etc. to be laid before the Directors of the Academy of Inverness.

Agreeable to the request of the Directors of the Academy of Inverness signified by James Grant

Esq. Provost of Inverness in his letters to Professor Playfair of the 22nd May last and 5th curt. We the undersigned met this day at the College Library to examine such candidates for the place of Rector in the Academy of Inverness as might come forward to submit their qualifications to a comparative trial. There appeared before us only Mr. Nimmo at present of the Academy at Fortrose who was accordingly examined by us on the subjects of Mathematics, Natural Philosophy, Chemistry, the Greek and Latin and French languages in all of which he acquitted himself very much to our satisfaction. In the sciences above named in which he underwent a long and particular examination he showed great accuracy and extent of knowledge as well as ingenuity and soundness of judgement. We can have no doubt therefore in recommending him the said Mr. Alexander Nimmo to the Directors as possessing the qualifications which they require in the Rector of their Academy in an eminent degree and as likely to discharge the duties of that office much to his own credit and the satisfaction of all concerned.

(Signed) Jo. Hill, Litt. Hum.
John Playfair, Prof. of Natural Phil.
James Finlayson, Prof. of Logic.
Thos. Chas. Hope, Prof. of Chemistry.

It will be seen that Nimmo was examined by four professors (not three as given in some sources), and for only one day (not three). John Hill was Professor of Humanity (Latin), although the signature does not fully indicate this. He retired about this time and died soon thereafter. Of the other professors, John Playfair is probably the one best remembered today. He was a mathematician, but in 1805 due to his interest in geology he changed his professorial chair from mathematics to natural philosophy. He was an original member of the Royal Society of Edinburgh. It is very likely that Nimmo had attended Playfair's mathematics lectures in the years that he was at Edinburgh University, and probably knew him even before this date when Playfair was tutoring Ferguson's sons in Kirkcaldy.

Nimmo was clearly the only candidate who went to the examination. As a result of this report, he was duly appointed. Pollock also had very good certificates, but had objected to the comparative trial by the professors, and was given expenses of £21 sterling. The directors decided that some 'grovelling' would not be out of order, and the following appears in their minutes:

> The meeting cannot avoid expressing the obligation the institution lies under to the gentlemen who have examined into and certified the qualifications of Mr Nimmo. It will be undoubtedly very creditable to the Inverness Academy to have the merits of its Rector ascertained by such eminent gentlemen and tho' the discharge of so useful and patriotic a duty is of itself sufficient recompense to persons of their sentiments yet the Directors in justice to themselves do request that Provost Grant will return to the different gentlemen through Mr. Playfair (who has occasion to take particular trouble in this business) their most sincere thanks

and acknowledgements for the services they have done to the Inverness Academy.[9]

Nimmo as Rector of Inverness Academy

Nimmo took up his duties either at the start of, or very early on in, the autumn session of 1805, the third teacher to come to the Royal Academy from the academy at Fortrose. He oversaw four other members of staff. On arrival he took over the teaching of the mathematics class, with class numbers varying from 17 to 23. From July 1808 he also taught geography, following the resignation of a member of staff. Numbers increased from 6 to 25 in the first of three six-month sessions that Nimmo taught this class, but then fell back.[10] In 1810 Nimmo was also teaching Spanish and Italian to a class of three pupils.[11]

A visitation committee had been appointed in December 1805 but no report of its work, nor that of a second visitation committee, was entered in the minutes book. The third visitation committee reported in March 1808.[12] Nimmo was not in his classroom at the time of the visitation, and was severely reprimanded for taking some of his pupils out to the field for practical surveying, when they should have been in the classroom. Although the directors did not doubt the value of this 'field work', they felt it should have been done in Nimmo's own time, so that the rest of the class could get the full lesson! The importance of staff and pupils attending church regularly on Sundays was also highlighted.

Soon after his arrival Nimmo had requested and been authorised to purchase globes, but his request to start a library was received with doubt by the directors, who referred the idea to a committee. (The local so-called 'Kirk Session' library was at this time housed in the Academy hall.) A tradition, which had continued from the Grammar School days, was the holding of 'orations' at the end of the June examinations, where pupils recited passages from famous authors, some in Latin and some in English. In 1808 an item was recited in Gaelic. In June 1808 and 1809, following the examinations and orations, a ball was held in the Northern Meeting Rooms in the town, but the newspaper reports do not indicate how much staff involvement there was in this.

In early 1811 Nimmo was suggesting a scheme of work for pupils to progress more consistently through the various subjects during their time in the school, rather than studying a single subject only to drop it.[13] The directors set up a sub-committee to consider these proposals but no formal report ever came before this committee, presumably as Nimmo's resignation was presented to the meeting on 11 June and the matter was dropped.[14] The January 1811 meeting had also requested the treasurer to enquire into the costs of repairs to the Rector's house carried out a couple of years previously, but at the May meeting the directors refused to fully fund the repairs—one of many occasions when they took this line—and the Deacon Convenor was told to claim the balance directly from Nimmo.[15]

The only recorded major incident involving Nimmo was in 1810. A letter was produced at a directors' meeting from Major Thomas Fraser of Newton, whose son Hugh was in the writing and arithmetic class, and also studying Latin:

> My son Hugh Fraser has attended the Academy for some years and I believe he behaved like other boys of his age and without giving much trouble to his Masters from disobedience or disrespect. It was therefore with much concern I learned that on Monday 29th ultimo he suffered a most cruel beating from Mr Nimmo, who, not satisfied with using his fists, seized a heavy mahogany ruler and with it beat the boy upon the shoulders, back and thigh. The marks and bruises are evidences of the usage he received which might, and still may prove of very bad consequences. However, I shall not at present enter into a more detailed account of the injury received, as that will best appear from the evidence of the medical gentlemen and the masters who saw the boy afterwards as well as the testimony of the other boys who were present.[16]

Nimmo was censured:

> The meeting having taken into consideration the facts admitted by Mr Nimmo himself respecting the manner and degree of chastisement inflicted upon Hugh Fraser are unanimously of opinion that he erred extremely in both, of the necessity of chastisement they are perfectly aware when necessary, but they are convinced it ought invariably to be inflicted with solemnity and a severity proportioned to the offence but in no case whatever with any other instrument than a tawse [*a leather strap*]. They therefore recommend to Mr Nimmo in future to beware of yielding to suddenness of provocation or hastiness of temper and if the

offence is of a serious nature to call the Masters and scholars of the whole Academy to the public hall and cause the master of the offending boy to inflict the chastisement there.

Nimmo as a surveyor and scientist.

During his time at both Fortrose and at Inverness Academies, Nimmo took an interest in various scientific matters outside the school, at a time when developments in arts, science and literature (later to be called the Scottish Enlightenment) were very much in vogue. Four pieces of evidence of his work at this time have been traced, and he also wrote a number of articles for *The Edinburgh encyclopaedia*, edited by David Brewster, who may have been a classmate of Nimmo's in Edinburgh. Some of these articles were completed after he moved to Ireland.

In December 1804 Nimmo was the author of a report presented at a meeting of the Royal Society of Edinburgh, although it was actually read to the meeting by Playfair. The title of the paper was 'An account of the removal of a large mass of stone to a considerable distance along the Murray Firth [*sic*]'. Unfortunately the report was never published, but from the title it is clear that Nimmo had been observing coastal erosion, transportation and deposition along the shores of the Moray Firth.[17] On 1 January 1811, shortly before he changed his occupation from schoolmaster to engineer, Nimmo was elected a Fellow of the Royal Society of Edinburgh, and in its register he is described, even at this early stage in 1811, as an 'engineer'. He was proposed as a Fellow by Sir George Steuart MacKenzie of

Coul (Ross-shire), a person with whom Nimmo had stayed in 1806.[18]

In the same year as he was observing and writing about sediment movement in the Moray Firth (1804), he was also making observations on Loch Ness in the company of his friend Simon Fraser of Foyers, Foyers being on the banks of the Loch. They were trying to discover the temperature of the water in the lower layers of Loch Ness, and lowered a half-gallon vessel to a depth of 120 fathoms (720 feet, or about 220 metres).[19] How they constructed the vessel to take the water samples is not known. During the experiment they drifted about a mile up the Loch even though the little wind that was blowing should have driven them in the opposite direction. They attributed this to a return current of water deep below the surface of the Loch. The first specific document that survives showing Nimmo's surveying interests is in the Highland Folk Museum collection at Kingussie.[20] It is a thumbnail sketch of Fort George, one of the key government forts built on the shores of the Moray Firth during and after the Jacobite Risings. A note on the reverse says 'Drawn by Mr Nimmo—in May 180?' It unfortunately is not clear what the last number is—it is either a 5 or 6, and has been changed from one to the other.

In the summer of 1806, during the school holiday from 6 June to 1 August, Nimmo surveyed the boundaries of the counties of Inverness, Ross, Sutherland, Caithness and Cromarty, and these were incorporated in Aaron Arrowsmith's map of Scotland, first published in 1807. He was paid £150 sterling for this work.

Nimmo's journal of this survey forms the basis of this present publication. Either in 1808 or 1809 he was to survey a route for a road from High Bridge (now Spean Bridge) in the Great Glen to Killin in Perthshire, via Rannoch Moor for which he was paid £50 by the Commissioners for Highland Roads and Bridges. The road was never constructed, despite being strongly favoured by Thomas Telford.

Nimmo's resignation

Joseph Mitchell, the civil engineer who learned his trade on the Caledonian Canal with Thomas Telford, and who later became Inspector of Highland Roads and Bridges, was living in Inverness just after Nimmo's time as Rector. Mitchell was a pupil at the Academy for three and a half years, from age 11, starting in about 1814, but no school record survives to confirm the details. His *Reminiscences*, however, do not seem to be very accurate.[21] After saying that Nimmo had established an admirable curriculum, Mitchell goes on to say:

> Students came from all parts of the surrounding country. About three hundred used to assemble in the Hall on Sundays to hear a prayer, and, headed by their masters in their gowns, they marched in procession to church. Nimmo, who was very social, used sometimes to neglect this duty on Sundays. The magistrates, austere men, particularly when their authority was questioned, censured him, and he resigned his appointment, to the great loss of the institution. He was immediately employed by Telford, and sent to Ireland as his assistant engineer.

The pupil numbers claimed by Mitchell are rather high, the highest recorded gross number (derived from the numbers attending each class) being 268 in 1810, while the actual numbers were probably below 250.[22] Nimmo may well have failed in his duty to lead the Academy to church regularly on Sundays (this had been hinted at in the 1808 visitation report), but his resignation was probably much more to do with the fact that the Commissioners for the Bogs of Ireland had appointed him as an engineer from 5 January 1811 for a period of three years.[23] He was required to work for 720 days in that period, at two guineas per day of actual employment. This would earn him a much greater sum than he was receiving as Rector of the Academy, which is unlikely to have been much more than £200 sterling adding together his salary and the relevant part of the fees paid by students. The 720 days could easily have been worked with a starting date late in 1811. It is likely that Nimmo resigned in order to take up this much better-paid job and pursue his scientific interests.

So, when Alexander Nimmo crossed the Irish Sea in the summer of 1811, Inverness Academy's loss became Ireland's gain.

Notes to Chapter II

[1] From the opening announcement for Inverness Royal Academy, in the form of an advertisement in the *Edinburgh Evening Courant*, 21 June 1792.

[2] Earl of Morton's Papers, National Archive of Scotland, GD150/3420 (Edinburgh, 1810).

[3] An early, but rather inaccurate, biography of Nimmo can be found in *A biographical dictionary of eminent Scotsmen, Volume V* (Glasgow, 1856); a much more

reliable one, but still with some minor errors, is in Ted Ruddock, *Biographical dictionary of civil engineers in Great Britain and Ireland, Volume 1: 1500–1830* (London, 2002), 483–9; the only full-length biography of Nimmo is by Noël P. Wilkins, *Alexander Nimmo, master engineer, 1783–1832: public works and civil surveys* (Dublin, 2009), on which some of this article has been based, especially the information about his early life and university career.

[4] R.N. Smart, *Biographical register of the University of St Andrews, 1747–1897* (St Andrews, 2004), quoted in Wilkins, *Alexander Nimmo.*

[5] Cupar Fife Town Council Minutes, 2 November 1796, quoted in Wilkins, *Alexander Nimmo.*

[6] Fortrose Academy was founded in 1791 but no detailed records survive from the first two or so years, thus the date of Nimmo's appointment cannot be checked; however, Wilkins has identified from the Presbytery of Chanonry session minutes that he subscribed to the precepts of the Presbyterian Church in February 1802, a condition of taking up the job.

[7] Inverness Academy Directors' Minutes, 2 April 1805.

[8] Directors' Minutes, 13 July 1805.

[9] Directors' Minutes, 13 July 1805.

[10] Staffing and pupil numbers extracted from the 'Rector's report of the state of the Inverness Academy to the directors, at their Annual Meeting on 30th April 1835', Inverness Royal Academy Archive, B2.

[11] An advertisement in the *Inverness Journal* on 8 June 1810 listed classes, class numbers and staff duties during the preceding session.

[12] National Archive of Scotland, GD128/34/1/36.

[13] Highland Council Archive, CI/5/8/9/3/1/1, letter from Alexander Nimmo to the directors, with inserted letter from Alexander Campbell, English teacher, not dated, but on the same topic.

[14] Directors' Minutes, 7 January 1811.

[15] Directors' Minutes, 20 May 1811.

[16] Directors' Minutes, 8 February 1810.

[17] J. Playfair in the Minutes of the Royal Society of Edinburgh (National Library of Scotland, Acc. 10000/4), as quoted in Wilkins, *Alexander Nimmo.*

[18] Noted by Wilkins, *Alexander Nimmo*, 25 and 41.

[19] Noted by Wilkins, *Alexander Nimmo*, from a paper by Nimmo 'On the application of the science of geology to the purposes of practical navigation', *Transactions of the Royal Irish Academy* XIV (1825), 39–50.

[20] Highland Folk Museum collection, Kingussie, ref. c.c.17; a note on a separate sheet of paper reads: 'Farr 23rd November 1864. The inclosed is a sketch of Fort George drawn by Mr. Nimmo Rector of the Inverness

Academy in May 1864. Which I had with me in India it is worn out by having been carried many years in my pocket book. AM'. The date attributed to the drawing on this cover sheet is clearly wrong, as Nimmo died in 1832, and it does not agree with the date given on the document itself. The writer of the note was Col Alex Mackintosh, and Farr is a community in Strathnairn to the south of Inverness, an area associated with the Mackintosh clan.

[21] Joseph Mitchell, *Reminiscences of my life in the Highlands, Volume 1* (Chilworth, 1883 and various modern editions).

[22] 'Rector's report of the state of the Inverness Academy'.

[23] Ruddock, *Biographical dictionary*.

Chapter III

The Scottish Highlands and Ireland in the Time of Alexander Nimmo

~ James Hunter ~

When, in the early summer of 1806, Alexander Nimmo set out on what he called his 'perambulation' of the borders of the Highland County of Inverness-shire, he was journeying into an area in the grip of a change so far-reaching as to be little short of revolutionary. For centuries, the Gaelic-speaking people of the Scottish Highlands—people whose language and culture derived from Irish immigrants who began moving into the region in the fourth or fifth century AD—had stood apart from their neighbours to the south. Elsewhere in Scotland and still more so in England, starting in the sixteenth century if not earlier, feudalism had gradually—or not so gradually—been giving way to an aggressively commercial capitalism. Throughout this period, however, the Highlands continued to be characterised by a much older mode of social organisation. Its embodiment was the clan, which, into the eighteenth century, retained key features of a tribalism long gone from most of the rest of Europe. Not the least significant of those features was the tendency of clan chiefs—MacDonald of Sleat, MacLeod of Dunvegan, Fraser of Lovat and many others—to take no very great account of the governments to which they were theoretically subject. Prior to the Anglo-Scottish Union of 1707, those governments were located in Edinburgh, Scotland's capital. From there, successive Scottish kings had sought to impose their will on the Highlands and, by so doing, bring the clans to heel. Some progress had been made towards this objective. But it had not been sufficient to ensure that the post-1707 United Kingdom, its political centre in faraway London, could take Highland loyalty for granted. On the contrary, there was a feeling among Gaelic-speaking Highlanders (just as there was similar feeling on the part of many Irish people) that it would be no bad thing if the United Kingdom government, together with the Hanoverian dynasty to whom ministers in London reported, were to be overthrown. This was with a view to reinstating the Stuart monarchy, which had been deprived of the thrones of England, Scotland and Ireland in the 1680s. To be pro-Stuart was, in the terminology of the time, to be a Jacobite. And because Highland clans retained their longstanding capacity to take armed action on their own account, those clans which clung to Jacobitism were—during the first half of the eighteenth century—key players in a series of uprisings or rebellions intended to bring about a Stuart restoration.

The most serious of the uprisings broke out in 1715 and 1745. After the first, London ministers—taking up where their Edinburgh-based predecessors had left off—set about pacifying the Highlands. They stationed troop detachments in barracks like the one at Ruthven in Badenoch, which Alexander Nimmo mentioned in the notes arising from his 1806 perambulation. They also initiated extensive road construction in what had previously been an almost road-less region. Nimmo, in 1806, made use of the resulting thoroughfares. But they were, to his mind, far from adequate—and he clearly had little time for General George Wade, the military man who had been placed in charge of post-1715 road building in Inverness-shire and other parts of the Highlands. 'General Wade was no engineer', Nimmo wrote.

That might be disputed. But what had long before become clear was that Wade's roads were a less than effective answer to the age-old problem of how to curtail Highland rebelliousness. When, in the summer of 1745, Prince Charles Edward Stuart landed on the western shores of Inverness-shire to launch a further Jacobite assault on the Hanoverian regime, the clan-based army which the prince put together was able—thanks to the government's laboriously constructed road network in the north—to march on Edinburgh at lightning speed, take the Scottish capital and crush the government force sent to Edinburgh's relief. Months later, Charles and his mainly Highland followers were at Derby, no more than 120 miles from London. This, to be sure, proved the highwater mark of Jacobitism, and Charles Edward Stuart's subsequent retreat was to end in April 1746 with his overwhelming defeat at Culloden on the outskirts of Inverness—the town which, 60 years later, contained Alexander Nimmo's home and workplace. But the spectacle of a Highland army within striking distance of London had so unsettled, indeed terrified, Hanoverian ministers that Culloden was followed by a concerted, well-resourced and ultimately successful drive to bring about the destruction of clanship. This, ministers correctly considered, was the basis of much of what made the Highlands so distinctive from, and so threatening to, the rest of Britain.

During the summer of 1746, the British army rampaged through the Highlands, hanging rebels, burning homes, taking possession of cattle and other goods. Tartan plaids or kilts—the traditional costume of Highlanders—were proscribed. The carrying of weapons, previously habitual, was outlawed. The legal and other powers exercised by clan chiefs were taken from them. Some Jacobite chiefs were executed, others fled abroad; most of them had their lands seized, or annexed, by the state. One outcome of this latter development was that all such annexed or 'forfeited' estates were mapped comprehensively by government surveyors. Hence the existence of those forfeited estate plans which Nimmo several times consulted in the course of his 1806 travels. As was equally true of similar mapping exercises in Ireland and in the United Kingdom's overseas colonies, the compilation of annexed estate plans was integral to the process, as government saw it, of taming localities that were previously, again as government had it, every bit as lawless and 'savage' (a word much applied to the Highlands in Culloden's aftermath) as they were unsurveyed. But if, in 1746 and for some years afterwards, the Highlands were thought in London to consist for the most part of inherently hostile territory, the region had long since ceased to be so considered when Nimmo left Inverness on his 1806 map-making mission. That it was so owed a great deal to the evolution of clan chiefs into a landed gentry which—in outlook, aspiration and language—was increasingly indistinguishable, by 1806, from landowners of the sort then to be found in every other part of Britain.

This development distinguished the Highland experience very sharply from that of Ireland—where Nimmo, still in his early twenties in 1806, would spend the greater part of his professional life. The Irish counterparts of Highland chieftains had mostly been removed from the scene in Elizabethan or Cromwellian times—and their lands parcelled out among incoming, usually English, landlords. It followed that the Irish landlord of the early nineteenth century, as far as his often Gaelic-speaking tenants were concerned, was an archetypally alien figure. In contrast, the Highland landlord of the same period, from the perspective of *his* mostly Gaelic-speaking tenants, was a man whose family predecessors, in their role as clan chiefs, had been regarded by the immediate ancestors of those same tenants as sources of the protection and security which tribal leaders have always been expected by their followers to provide. The landlords Nimmo encountered in the Highlands might have managed—indeed did manage—their estates in much the same way, and with much the same objectives in mind, as those other landlords he would afterwards meet in Ireland. But for all that Irish and Highland estate management policies approximated to each other, and for all that those policies often impacted very negatively on estate tenantries, residents of Highland and Irish estates tended to respond in very different ways to what was done to them by their social superiors. The Irish landlord, seen widely as in every respect a foreigner, was an easy man for his tenants to hate and to resist—sometimes violently. The Highland landlord, though condemned in Gaelic song and poetry for having turned his back on clanship and the wider culture with which it had been associated, was a lot less likely to encounter active opposition from people who still felt him, however anachronistically, to be one of them. Hence the embittered comment made by Hugh Miller, a Highland-born journalist and commentator when, in the 1840s, famine came to both Ireland and the Highlands: 'They [the Irish] are buying guns, and will be by-and-bye shooting magistrates and clergymen by the score; and Parliament will in consequence do a great deal for them. But the poor Highlanders will shoot no-one...and so they will be left to perish unregarded in their hovels'.[1]

In the event, Ireland's famine proved enormously more catastrophic than its Highland counterpart. Miller, to that extent, was wrong. But he had all the same put his finger on a key distinction between the Highlands and Ireland. After 1746 at any rate, there was to be far less rebelliousness in the former area than in the latter. Would this Highland quiescence have been less evident if, as was contemplated briefly in London, the British government had responded to Charles Edward Stuart's uprising by following Irish precedent and expropriating Highland chiefs in order to replace them with landlords imported from the south? Nobody can answer that question with any certainty. What is beyond doubt, however, is that it was to be a long time before the generality of Highlanders shed the residual loyalty felt towards chiefs-turned-landlords and embarked on an Irish-style protest of their own. This was despite the fact that, during the decades following the Battle of Culloden, people throughout the rural Highlands were treated by the region's increasingly anglicised ruling order in ways that were more and more overtly exploitative.

The beginnings of commercial land management in the Highlands are evident well before 1746. But the trend towards the adoption of such management was accelerated by developments which the British government did a great deal to foster and promote in the post-Culloden era—not least the tendency, on the part even of formerly Jacobite chiefs or their sons, to adopt the lifestyles of the southern aristocrats in whose company they now spent much of their time. Gaelic-speaking clan chiefs had always been immensely status-conscious—as is made clear by the poetic tributes composed in their honour by clan bards. This remained true of their English-speaking successors. But while the standing of earlier MacDonalds of Sleat or Frasers of Lovat had depended on the number of fighting men at their command, the prestige and position of such men's late-eighteenth-century descendants depended much more on their spending power. 'The number and bravery of their followers no longer supports their grandeur', a late-eighteenth-century Highlander commented of his area's upper class. 'The number and weight of their guineas only are put in the scale'.[2]

Where landholding had previously been organised with a view to maintaining the clan as a war-making machine, it was now organised in ways intended to maximise an estate's revenue-generating potential. From the 1750s forwards, it followed, rents were everywhere jacked up. To begin with, those rents were paid out of the proceeds of cattle-rearing—a longstanding Highland activity. During the closing years of the eighteenth century, however, and even more especially in the course of the succeeding century's opening decades, cattle were replaced steadily by sheep—with the aim of enabling Highland landlords to cash in on soaring demand for the wool needed to clothe the burgeoning populations of Britain's rapidly expanding urban centres. 'In this as in every other instance of political economy', remarked an adviser to one of the leading landed families in the Highlands, 'the interests of the individual and the prosperity of the state went hand in hand. And the demand for the raw material of wool by the English manufacturers enabled the Highland proprietor to let his lands for quadruple the amount they ever before produced to him'.[3]

Cattle had traditionally been reared by the indigenous population of the Highlands, who used methods which Nimmo commented on in the course of his 1806 travels through the region. 'The tenants', he wrote of the inhabitants of the typical Highland township, 'kept as many cattle as they could winter on...their arable land'. With the approach of summer those cattle—together with spring-born calves which would eventually be driven south for sale—were moved high into the hills where they grazed on seasonally productive upland pastures and where, as Nimmo remarked, their herders 'dwelt in small temporary huts called Bothies or Shealings'. There, cattle remained until summer's end—when they were returned to lower-lying arable land which had been cropped in their absence and where permanent human settlement was concentrated. So things had been for generations. But for all its longevity, virtually nothing of this pattern of land-use survived the introduction of large-scale sheep farming. Such farming required capital, as well as skills, of a sort few Highlanders possessed. And the incoming farmers who consequently dominated the business—and who imported shepherds as well as sheep—had no more use for the cattle-rearers they replaced than they had for the former group's livestock. 'Hence we may account for that species of policy which induces the sheep farmer to remove as many small tenants as he can', Nimmo wrote in 1806.[4] Hill grazing given over traditionally to cattle was now monopolised by new men's sheep. Former arable land—being 'green' because of its 'having been long under the plough', as Nimmo observed—was taken over by those same new men in order to provide their flocks with winter pasture. In consequence, whole communities ceased to exist—hundreds, indeed thousands, of families being evicted by their landlords in the course of the mass removals which became known as the Highland Clearances and which culminated, very often, in a single sheep farmer and his shepherds occupying land which had previously sustained the inhabitants of a large number of townships.

The results of those upheavals were apparent almost everywhere Alexander Nimmo went in the course of his 1806 perambulation. Noting the 'immense torrents of loose gravel and stones' to be seen in Glencoe, Nimmo—in an early observation of the ecologically disastrous effects of the over-grazing that became commonplace in the Highlands during the nineteenth century—commented that 'these land slips [had] increased much' since the commencement of sheep farming. 'On purpose to improve the pasture', he wrote, 'the heather and young wood [,] whose roots bound the loose soil of the hills [,] have all been burnt off and eradicated while the sheep form tracks and roads on the sides of the mountain, which serving to convey the water of a considerable tract to one point, form falls that cut upon the loose soil till at length the bank slips down and sets many tons of this loose friable stuff into motion'.[5] Extensive heather-burning of the type thus criticised by Nimmo began with the introduction of sheep farming. It has continued down to the present day—and has all too often had consequences of exactly the sort Nimmo was among the first to describe.

One other outcome of sheep farming's growing prevalence in the Highlands of 1806 was to impinge directly on the job that Nimmo was then doing. Boundaries of the type he had been hired to map had not mattered greatly in the Highlands of the past. Then rents were levied, principally on small and readily identified areas of arable land. As a sheep economy displaced the cattle and cropping economy of earlier times, however, the extent of a farm's hill grazing became a key factor in its value. Hence Nimmo's 1806 remark to the effect that, on 'lands [having] been rented with a view to the pasture alone[,]...it became necessary for the proprietors to have their marches [meaning boundaries] more distinctly settled'.[6]

Fixing a boundary in this new fashion meant getting it down on a paper-based map—as opposed to its location being one of many items of customarily-acquired knowledge transmitted orally from one generation of Gaelic-speaking Highlanders to the next. All such knowledge, after all, was being lost along with those communities which had fallen victim to clearances—clearances which surface here and there in Nimmo's narrative. Writing of his visit to Lord Lovat at the latter's Beaufort Castle home some ten miles north-west of Inverness, Nimmo describes a large-scale map he was shown there by Lovat himself. This map—'executed with great care', in Nimmo's judgment—was one of those prepared on the orders of the Forfeited Estates Commission half a century before. Its subject area was Glenstrathfarrar—some miles further to the north-west and a district where clearances had got underway not long before Nimmo embarked on his own survey of the same locality. 'Several places are marked on the [Forfeited Estates Commission] map...which by the introduction of sheep are now uninhabited',[7] Nimmo reported of Glenstrathfarrar.

This was in 1806. Clearances on an extensive scale—one that became still more extensive with time—would continue in the Highlands for another half century. Initially, neither the landlords who ordered mass evictions nor the government which endorsed their right to do so were in any way intent on depopulating the Highlands. Estate-owners, for their part, were still inclined to treat evicted families as having a potential contribution to make to the expansion of estate economies. In the eastern and central Highlands, some such families were moved into villages like Kingussie, a settlement which Alexander Nimmo thought, over-optimistically as it turned out, to be 'well adapted for an inland manufacturing town'.[8] In the west, meanwhile, thousands of evicted Highlanders—people who had originated in the more inland areas then being given over to sheep—were settled on newly-created smallholdings or crofts. These typically consisted of three, four or five acres of indifferent arable land—the objective of the landlords who laid them out being to force crofters, as the occupiers of such holdings were called, to become either fishermen or labourers in the so-called kelp industry. This latter enterprise—which also flourished on the west coast of Ireland where Nimmo would encounter it in the course of his first foray into Connemara in 1813—involved the manufacture from seaweed of a crude industrial alkali then much in demand by manufacturers of soap and glass. Because landlords profited even more from kelp than from sheep farming, they were—at the time of Nimmo's Inverness-shire perambulation which coincided with the height of the kelp boom—prospering mightily. All across the Highlands, newly installed sheep farmers were paying higher and higher rents—while the many victims of eviction, in their role as kelp-manufacturing crofters, were contributing just as substantially to estate revenues. The era, it followed, was one of some optimism about Highland economic prospects—on the part, at least, of the people who were both bringing about and benefiting from the region's still-accelerating transformation. Hence, perhaps, the extent to which Nimmo's 1806 notes are replete with comments to the effect that, given a new road here or some bridges there, the north of Scotland could not help but flourish.

This did not happen. By the 1820s, with the invention of chemical processes capable of turning out endless quantities of cheaply produced alkali, the kelp boom was over. Crofters, as a result, began to be regarded by their landlords as a 'redundant' population. And British governments, which had formerly wanted to retain population in the Highlands, if only because the region had become a key supplier of military manpower in the course of Britain's wars with revolutionary and Napoleonic France, began to agree with the growing number of landlords who now aimed to rid the Highlands of as many people as possible. Government's former commitment to helping expand the Highland economy through the provision of canals, roads and other infrastructure—a commitment which had brought about the need for the county boundaries survey on which Nimmo was employed in 1806—gave way to a growing insistence that the only solution to

Highland difficulties lay in reducing the overall population of the area by means of encouraging emigration.

The notion of quitting the Highlands for North America—which ordinary Highlanders, like their Irish counterparts, regarded as a place of opportunity—was one that made some sense to crofters. But having by this point been thrown back, in the absence of their former employment in the kelp industry, on meagre agricultural resources, most crofters could not afford the cost of a transatlantic passage. As population continued to expand and as congestion grew on the comparatively small areas of land available to the crofting population, croft after croft was given over entirely to potatoes—the only crop, because of the diminutive nature of his croft, which the typical crofter could grow in sufficient quantity and that would come close to supplying the needs of his family. This universal reliance on a single, highly vulnerable source of nutrition—when combined with the increasing impoverishment resulting from an almost total absence of worthwhile paid employment—meant that, by the 1820s and 1830s, the crofting districts of the Scottish Highlands were beginning to resemble those Connemara localities where, as Noël Wilkins comments in his account of Alexander Nimmo's time in Ireland, Nimmo from an early stage encountered 'scenes...of utter desolation and human misery'.[9]

When, in the 1840s, potatoes in the Highlands—as in Ireland—were blighted year after year, the immediate outcome was famine. A secondary outcome was a renewed wave of clearance and eviction—accompanied this time by often enforced and sometimes subsidised emigration to Canada and Australia. Because blight came to the Highlands a year later than it did to Ireland, because government was thus more prepared to deal with its consequences and because the affected population—some scores of thousands in the Highland case as opposed to millions in the Irish one—was much smaller, what happened in the Highlands in the middle years of the nineteenth century, though bad enough, is not to be compared with the tragedy then unfolding in Ireland. The two famines, for all that, had similar causes and affected people of similar cultural background—this background, then beginning to be widely described as Celtic, being one which mainstream opinion in Britain was increasingly prone to regard with contempt. 'It is a fact', a southern journalist observed of Highlanders during the famine years, 'that morally and intellectually they are an inferior race...[T]he real cause of their destitution is the idleness that is rooted in their very nature'.[10] When in Ireland, it is almost needless to say, Alexander Nimmo heard much to the same effect. He, however, was not of this mind. 'Really', Nimmo told a House of Lords committee who had put it to him in 1825 that Irish indolence was the source of Irish troubles, 'I am not inclined to think that the Irish are an indolent people...I have a far higher opinion of the spirit of independence of the people than perhaps many persons who are immediately concerned with the country'.[11]

As indicated by his account of his 1806 perambulation of Inverness-shire, and as demonstrated by the Irish surveys he undertook subsequently on behalf of various official agencies, Nimmo was always at pains not only to get right the spelling of Gaelic place names but to come to some understanding of what those names signify. His interest in the shared Gaelic language of Ireland and the Highlands, together with his hostility to anti-Celtic feeling of the kind that caused him to take issue with the Lords committee of 1825, might have led Nimmo—or so one can speculate—to have some sympathy with the political alliance that developed between Irish and Highland land reformers in the later nineteenth century. This alliance reached its high point in the 1880s when, borrowing techniques first developed by the Irish Land League, crofters at last mounted effective resistance to policies of the sort that had produced the Highland Clearances. In the Highlands during the 1880s, rents were withheld, sheep farms were occupied illegally, a Highland Land League was organised and pro-crofter MPs were elected to parliament. Although the British government made some attempt to safeguard the *status quo* by means of troop deployments and other measures of that kind, substantial concessions were eventually granted. Of those, the most significant was the Crofters Act of 1886 which, because it provided crofters with security of tenure, made further clearance impossible. In the course of their battles for this legislation, crofters had strong support from Ireland. A major source of such support was the Irish Land League's founder, Michael Davitt, who in 1887 made his own 'perambulation' of the Highlands where, on the Isle of Skye, a Highland Land League stronghold, this nationalist and Catholic Irishman was fêted by thousands of staunchly Presbyterian crofters.

Michael Davitt had been born in the famine year of 1846 in Co. Mayo. More than a century after his triumphant trip to Skye, the Inverness-shire island received a visit from another Mayo-born person of consequence. This was Mary Robinson, then president of Ireland. 'I'm very conscious', Mary Robinson said on that occasion, 'that in coming to Skye...I have followed in the footsteps of a great Irishman, Michael Davitt... Growing up as a child in Mayo, I learned so much about Michael Davitt and admired so much his commitment...to reform of the land and the founding of the Land League...You may recall that he received a wonderful welcome here in Skye in May 1887 because he had championed the cause of Skye crofters...These are common experiences which mark forever the history and development of our nations...I hope that, in considering our past, we can recognise what we have in common and cherish that'.[12]

Among much that is common to Ireland and the Scottish Highlands—and needing, therefore, to be remembered in the way Mary Robinson suggested—is the career of Alexander Nimmo. From his Inverness-shire beginnings as a map-maker, he would go on to become one of the foremost surveyors and engineers in the history of Ireland—where, incidentally, Nimmo's many infrastructural achievements include still-surviving Mayo roads and bridges that would have been as familiar to Michael Davitt as they would afterwards be familiar to the young Mary Robinson.

Notes to Chapter III

[1] W.M. MacKenzie, *Hugh Miller: a critical study* (London, 1905), 190–1.
[2] James Hunter, *The making of the crofting community* (Edinburgh, 1976), 11.
[3] J. Loch, *An account of the improvements on the estates of the Marquess of Stafford* (London, 1820), xvii.
[4] Alexander Nimmo, 'Journal along the North East and South of Inverness-shire. Ends at Fort William', f. 39.
[5] Nimmo, 'Journal', f. 42–44.
[6] Nimmo, 'Journal', f. 19.
[7] Nimmo, 'Journal', f. 7.
[8] Nimmo, 'Journal', f. 28.
[9] Noël. P. Wilkins, *Alexander Nimmo, master engineer, 1783–1832: public works and civil surveys* (Dublin, 2009), 187.
[10] K. Fenyo, *Contempt, sympathy and romance: Lowland perceptions of the Highlands and the clearances during the famine years, 1845-1855* (East Linton, 2000), 61–2.
[11] Wilkins, *Alexander Nimmo*, 326.
[12] Mary Robinson, President of Ireland, 'Oraid sabhal mòr ostaig', lecture (Skye, Glasgow, 1997), 12.

Chapter IV

A Cartographic Perambulation Around Alexander Nimmo's Inverness-shire Journal

~ Christopher Fleet[I] ~

From a cartographic perspective, Alexander Nimmo's journal provides an informative, first-hand account of several subjects. It is a useful source of information regarding the construction of one of Scotland's most important maps, Aaron Arrowsmith's *Map of Scotland* of 1807. Nimmo's account confirms the thoroughness and care taken in the compilation of this map, as well as providing a helpful evaluation of its trustworthiness for different categories of information. His journal is an honest and perceptive assessment of the varied utility of different sources of information—maps, charters, shepherds, tenant farmers, estate factors, and landowners—in confirming the complex patterns of land ownership for farms, landed estates and county jurisdictions. Nimmo also provides an engaging narrative of the practicalities of map and boundary delineation in the early nineteenth century. Through his comments on mapping and surveying, he illustrates the changing nature of map-making and the values that both drove it and were reflected in it at a key moment in time.

This essay discusses these issues and offers some background and context to Nimmo's Inverness-shire journal. First, the rationale for Aaron Arrowsmith's Scotland map, its commissioning and construction, and Nimmo's role in it, are described. Second, the various other maps of Scotland that were used by Nimmo and mentioned in his journal are reviewed. A final section brings these threads together by looking at the changing social and cultural values reflected in Nimmo's journal and his maps.

Aaron Arrowsmith's Map of Scotland (1807)

The 1803 act establishing the Parliamentary Commission for making Roads and building Bridges in the Highlands of Scotland had laid down that half of the costs of such works would be paid by proprietors; very quickly, however, disputes arose over allocating the share of these costs. One of the Commissioners' earliest planned roads was the Glenshiel Road, leading from Invermoriston by Loch Ness over to Loch Duich and the ferries to Skye. The route led through the counties of Inverness-shire and Ross-shire, but huge difficulties arose over allocating the potential costs between these counties. The more detailed extant surveys of estates showed inconsistent or ambiguous boundaries, while, in contrast to southern Scotland, no county maps had been published by this time for those counties north of the Great Glen.[2] The Commissioners' solution to this problem was to sponsor a new map of Scotland, one based on the best available information and the latest mathematical and scientific principles, which would show county boundaries with precision and which would be respected as an authority for allocating the financing of road and bridge construction work. They turned to Aaron Arrowsmith to produce this map, and to Alexander Nimmo to define the northern county boundaries.

Aaron Arrowsmith (1750–1823) was born in Co. Durham in 1750, and worked as a surveyor before setting himself up as a map publisher in London from 1790. His extensive output included general and historical atlases of the world, geography text books and topographic maps, many of which went into authoritive editions. While he himself undertook no direct surveying work, he gained a reputation for the quality of research behind his maps and for incorporating new information from the latest discoveries. Upon learning of Arrowsmith's work on a new and detailed map of England,[3] the Commissioners felt that Arrowsmith should be invited to do the same for Scotland.

This was an age that encouraged the publication of maps with accompanying memoirs, describing the compilation of the map and its sources and acting as an important testament to the trustworthiness and quality of the map itself. Arrowsmith's *Memoir relative to the construction of the map of Scotland* is a minor classic of its kind, packed with useful and insightful information.[4] Arrowsmith consulted over 100 maps and plans as source material for the map, including nautical charts, estate surveys, county maps and road surveys, which he carefully listed in an appendix. His informants included scientists, landlords, academics, ministers, teachers, as well as engineers in the new Ordnance or Trigonometrical Survey, all personally credited and each helping lend authority to the map. Nimmo is credited several times in the *Memoir,* not only for his Inverness-shire survey, but also for his assistance in correcting place-names, in providing a sketch map of the Black Isle, and for his description of the Monadhliath Mountains between the Great Glen and Speyside. But Arrowsmith's greatest cartographic debt was to William Roy's Military Survey of Scotland (1747–55) which, through the assistance of John Rickman, Arrowsmith was able to trace onto transparent paper with his assistants in the King's Library between August 1805 and March 1806.

Roy's Military Survey, often referred to by contemporaries as 'The Great Map', was the most detailed and comprehensive survey of mainland Scotland in the eighteenth century. Drafted as a direct consequence of the Jacobite Rebellion of 1745–6, when military commanders found themselves 'greatly embarrassed for want of a proper Survey of the Country',[5] it brought together the best of contemporary military engineering and draftsmanship. Whilst Roy considered the work more 'a magnificent military sketch than a very accurate map of a country',[6] it was the first time teams of engineers had worked collectively to map Scotland, and the first time that theodolites (precise measuring tools for more accurate positioning) and chains had been so widely used, albeit only for selected landscape features. Yet, for nearly half a century, Roy's map had lain hidden from public view in the King's Library in London, and the mapping of Scotland had been necessarily based on less detailed and accurate sources. As Arrowsmith well knew, the major strength of his own map lay in discovering and making available the unrivalled detail and quality of Roy's Great Map.[7]

However, the Military Survey had been made for a very different purpose, and its strengths and omissions to Arrowsmith were also highlighted by Alexander Nimmo's survey. Features of interest to an army commander—roads, rivers, the general position of villages and hamlets, designed landscapes around the larger country houses, as well as land-cover and terrain—were often remarkably accurate. But Roy's military engineers had no interest in property or administrative boundaries, in naming or recording smaller topographic features, or in recording the correct form of place names, a problem compounded by their general unfamiliarity with Gaelic. Although in later years Roy himself became a celebrated geodesist and laid the trigonometric basis for what became the Ordnance Survey, his Military Survey lacked an accurate scale or the inclusion of a graticule (a neat, geometrical grid of lines of latitude and longitude), omissions which were increasingly seen as problems by Arrowsmith's time. Finally, Roy's survey was over 50 years old by the time Arrowsmith worked with it, and given the half-century of changes which took place, such as rural depopulation, emigration, urbanisation, landscape improvement and the arrival of new roads and even canals — much of which is detailed in both James Hunter and Noël Wilkins' essays in this volume — considerable updating was required.

Fortunately, Nimmo took a broad interpretation of his remit, and he was able to help remedy all of these problems. We learn from the journal that he received Arrowsmith's draft map on 20 May 1806. He set to work the next day, visiting Beaufort Castle and examining maps of the Lovat estate showing the northern boundary of the county (see map). A visit to Dingwall on 3–4 June allowed him to examine plans for the forfeited estate of Cromarty, and the north-eastern boundary by Strathpeffer. He then proceeded in a generally clockwise direction along the eastern and southern parts of the county, bordering successively with the counties of Nairn, Moray, Perth and Argyll, taking the new Loch na Gaul road to Arisaig. He then took a boat to Kintail and ascertained the northern county borders with Ross and Cromarty, and with Sutherland, before returning to Inverness by 1 August.

By 12 October 1806 he was able to send Arrowsmith a mass of information on northern Scotland, beyond his main remit of delineating the county boundary on the draft map. His 'Historical statement of the erection and boundaries of the northern shires' was also transmitted to Arrowsmith at this time, and is perhaps the most thorough description ever made of the historical background and authoritative sources for these county boundaries.[8] He sent Arrowsmith larger-scale sketch maps (two of which are reproduced here from his journal) where space did not allow proper corrections to the draft map. This was particularly necessary in the Black Isle, of which he wrote that it 'affords a specimen of political geography that I think may justly vie with any tract in the German Empire for intricacy'.[9] Nimmo was the first to record the 14 detached districts of Ross-shire, scattered irregularly throughout Cromarty and until then ignored by map-makers; through this, the old shire of Cromarty was augmented fifteen-fold. His journal gives ample evidence of his keen interest in the etymology of place names, and he was able to correct the orthography of many names on Arrowsmith's map. He also recorded new names of smaller farms and shielings, and deleted those that had been abandoned. Although not required for positioning the boundary, he took celestial observations where possible, understanding the importance of greater positional accuracy to Arrowsmith. He even corrected coastlines—for the coast of Easter Ross he 'marked an outline that appears pretty much the truth'—and the shapes and sizes of lochs: 'I fear Loch Levin is (on the large map) made rather too long'.[10] In short, Arrowsmith had every reason to be grateful to Nimmo, whose work also helped to endorse Arrowsmith's method:

> Mr Nimmo also transmitted a journal of his proceedings, which was of great use to me; besides that the whole of his perambulation really and necessarily formed a severe scrutiny of the accuracy of the Military Survey, which was found more correct, except in a few trifling instances, than was expected from General Roy's account of it.[11]

Estate surveys, county mapping and road surveys

As Arrowsmith and Nimmo well knew, in spite of the accuracy of the Military Survey, Scotland's geography was more interesting and complicated than Roy had shown it to be, and it was necessary to consult other sources. These included estate maps, county maps and road surveys, all of which were found useful but, in some respects, still wanting. Historic charters and related legal documents were also consulted. First-hand information was provided directly by the spoken word of shepherds, factors and tenant farmers, whose testimony was often more honest and useful than that of their paymasters and landlords.

Nimmo was able to benefit from the mapping funded by the Commission for the Forfeited Annexed Estates. This included the work of Peter May on the Lovat estates and the estate of Cromarty, both on the northern boundaries of Inverness-shire; as well as the work of William Morison on the estate of Lochiel in the west.[12] The Commission had administered 13 pro-Jacobite estates following the Annexing Act of 1752 until 1784. Detailed surveys were ordered, and suggestions carried into effect for a wide range of agricultural, social, educational and infrastructural improvements, as well as encouraging English and banishing Highland dress. There was a military connection to this mapping, since Lt Col David Watson, who had funded and promoted Roy's Military Survey of Scotland, instructed surveyors and was very influential from the outset as a Commissioner.

Estate mapping, funded by landlords and executed by private surveyors, provided a more extensive source of boundary information. Nimmo was able to utilise these maps best where the county boundary traversed more lowland areas, for example, at Culloden, Kilravock,

Cantray, Castle Grant, Glen Tromie, and Fassifern. There was rapid growth in estate mapping in parts of the Highlands following the cessation of Jacobite hostilities. Difficult years of agricultural depression in the 1780s were followed by a further boost caused by the disruption to trade during the Napoleonic Wars from 1793 to 1815, and the resulting high prices for agricultural commodities.[13] Nimmo made good use of these maps: the landscape they portrayed was often viewed through the eyes of an improver, recommending or proposing new developments. Most usefully too, they were detailed and large in scale, often surveyed using trigonometry, and with lines indicating ownership.

Estate surveyors often turned their hand to smaller-scale maps of counties. This was true for all three of the county map-makers whose work Nimmo mentions: James Stobie, George Langlands and John Ainslie. From 1759 the Society for the Encouragement of Arts, Manufactures and Commerce (from 1847, the Society of Arts) had promoted county mapping by offering awards of up to £105 for those that were based on original surveys, triangulation and accurate measurements of latitude and longitude, as well as drafted at a scale of one inch to the mile or larger.[14] This particularly encouraged county maps in the Scottish borders, where a larger base of subscribers could fund the work, but greater levels of patronage from landed society was necessary to extend this mapping to the more sparsely-populated northern counties. James Stobie, who worked as factor to the 4th Duke of Atholl, used all the influence and connections of the Duke and his brother in law, Thomas Graham of Balgowan, to finance his magnificent county map of Perthshire and Clackmannanshire, published in 1783.[15] George Langlands, who worked with his son Alexander as an estate factor and agricultural advisor for the 5th Duke of Argyll, was supported by the Duke for his county map of Argyll of 1801.[16] It is interesting to read documentary evidence from Nimmo of these maps in use in quite remote quarters: Stobie's map in a shooting lodge by the shore of Loch Ericht, and Langlands' map in a farmhouse in Glencoe. John Ainslie (1745–1828) was one of the most outstanding of Scotland's mapmakers. He was supported by commissions from many landlords, as well as by the Commission for the Annexed Forfeited Estates. He was also a founding member of the Society of Antiquaries of Scotland, and used the Society as a setting to negotiate and secure patronage.[17] His 1789 map 'Scotland, drawn and engraved from a series of angles and astronomical observations' was the standard map of the country for two decades before Arrowsmith, and was a major advance on earlier maps by men such as James Dorret.[18] Ainslie also trained the land surveyor William Bald who, like Nimmo, would go on to work on the Bogs Commission maps in Ireland.

Whilst surveys of roads are only mentioned in passing in Nimmo's journal as topographic evidence, he was clearly keen to promote new routes to John Rickman and the Commissioners, and this was an active period in Highland road-surveying. During the eighteenth century, the planning of new roads and canals was increasingly associated with maps, but by different practitioners for different purposes. Military imperatives under Wade from the 1720s, and his successor Caulfeild from the 1740s, had resulted in the construction of over 1,050 miles of roads in the Highlands by the 1760s; these maps share many characteristics of other military maps.[19] Thereafter, road construction was steadily devolved to landed and county interests, with maps often made by estate surveyors and civil engineers. Between 1790 and 1799, George Brown, a civil engineer and estate factor based in Elgin, surveyed over a thousand miles of roads across Sutherland, Caithness, Ross-shire and Inverness-shire on behalf of the British Fisheries Society and the Highland Society. Brown's road maps were used by Arrowsmith, and his surveys and work were carried forward by the Parliamentary Commission for making Roads and building Bridges in the Highlands of Scotland. During the 1790s Thomas Telford had also acted as a surveyor and engineer to the British Fisheries Society, providing an essential basis for his work from 1803 with both the Commission for roads and bridges, and that for the Caledonian Canal.

Although Nimmo does not refer directly to maps dividing common land or 'commonties', his journal provides interesting remarks on these and observations of his own role in carving boundaries through them. Historically, commonties had played a vital part in subsistence agriculture for grazing, fuel, building materials and reserve land to cope with population growth. However, following acts of parliament in the 1690s, this land was increasingly regarded as economically useful and commonties were divided, often by use of a special surveyed map. It has been estimated that half a million acres of commonties were divided in this manner between 1695 and 1900. In several areas that Nimmo passed through, these commonty divisions were very recent—in Strath Spey near Grantown and at Aviemore, at Drumochter, and above Loch Laidon. In these areas the marches across commonties formed the county boundary, and were often vague and sometimes disputed: 'In other places the boundary is only known with certainty in the arable lands, the upper country being pastured in common, and as the whole belonged to one proprietor, the limits of the county have never been defined.'[20] Arrowsmith's map is the first known record of these boundaries on the ground, and Nimmo's journal and his 'Historical statement' explain the reasoning behind them.

At Drumochter and Loch Laidon, boundary disputes form a wider interesting backdrop to Nimmo's journal. From the dilatory and even hostile responses Nimmo received from the County Commissioners for Inverness-shire and Perthshire, it was clear that several powerful landlords were not in favour of his boundary survey, and were aware of the new expenses it would usher in. This was already a familiar problem for the Commissioners. George Mackenzie of Coul, Convenor of the County of Ross, was described by John Rickman in 1809 as 'a man of some genius, but not of the kind which facilitates business'.[21] Nimmo too conceded to the repeated requests from George Mackenzie to view Arrowsmith's draft map, but sensibly chose to decline his distracting suggestion to measure a more accurate base in the Black Isle for a fresh survey. Further south other landlords such as the Duke of Gordon failed completely to reply to Nimmo's letters, causing much extra work on the ground, whilst at Rannoch, the rival parties in a land dispute stayed away, perhaps fearing that Nimmo had the authority to determine their own estate boundaries as well as just the county boundary. Behind the practicalities of defining the boundary, power was shifting away from landlords to new county and government officials. Some landlords had more to lose than others through Nimmo's work.

Enlightenment mapping and values

Nimmo's journal provides a refreshing and quite rare evaluation of contemporary mapping by a user in the field. Before the Ordnance Survey's work in Scotland in the nineteenth century, mapping was largely funded from private sources, with only occasional government sponsorship. It lacked the systematic and comprehensive basis that Arrowsmith and Nimmo sought to achieve. But it also reflected different values. The maps that Nimmo consulted—especially estate maps—were driven by and sometimes reflected overtly political purposes. For the forfeited estates, these purposes included the destruction of Gaelic and clan-based customs and practices, whilst more generally, estate maps were key elements in agricultural reorganisation and improvement, in commodifying the landscape, and in encouraging socio-economic change towards a cash-based economy. Perhaps Nimmo would have seen natural benefits to furthering these purposes—especially in his future work in Ireland—and his comments on the maps tend to be focused on the positional accuracy of their boundaries and features for his own survey.

In these concerns for accuracy and precision, and in his often critical remarks about earlier maps, Nimmo exemplifies and illustrates the broader ways in which maps were being seen and used at this time. During the Enlightenment, accuracy of measurement came to be regarded 'as the *sine qua non* of cartographic progress', and there was also 'an increasing emphasis in mapping on original survey, on more precise instruments, especially at sea, and on more detailed cartographic representation as an end in itself'.[22] Under this prevailing *esprit geometrique,* maps became plainer, stripped of artistic embellishments, and assembled an expanding range of geographical 'facts' within a precise and coherent graticule. To an extent we can see this transition, from the Military Survey through to the work of Ainslie and other county map-makers, to Arrowsmith and the Ordnance Survey, but this would have been felt more strongly by early-nineteenth-century observers such as Nimmo and Arrowsmith. The Enlightenment also encouraged the belief, still widely held today, that maps progressively shed error in a steady improvement in the way the world is represented. The belief in the map as an 'objective' or 'truthful' representation of reality can be traced to this time. In practice, neither of these claims can be supported without qualification, and a focus on the context of the maps, their makers and users, and on the social and political purposes behind their making, can allow a broader understanding of the maps themselves.

This social and intellectual context is useful, nevertheless, in understanding Nimmo's own value system and purpose with regards to his mapping and survey work. His interest in accurate instruments of measurement, celestial observations, scientific rigour, direct survey on the ground, and the need to locate and position places and features with authority and without ambiguity are recurring themes. So too was his need to criticise earlier sources and maps on these new terms. Nimmo describes Stobie and Langlands' county maps, as well as Ainslie's 1789 map of Scotland in quite scathing terms, 'not to be compared in point of accuracy or minuteness with the skeleton map I have'.[23] This was, of course, a rhetorical claim, especially given that his recipient was the author of the skeleton map. But as a claim it illustrates his strong need to judge the new as better than its predecessors. Similarly, Arrowsmith's remark that Ainslie 'does not appear to have possessed the means of improving the delineation of the interior of the northern counties' was a fair criticism, and provided something of the rationale for his own map and Nimmo's work. In areas so clearly wanting better maps, direct evidence from authoritative people 'out there' was also highly valued, and Arrowsmith enjoyed relating, in his *Memoir,* that Mr MacPherson of Ralia was 'a venerable old gentleman eighty years of age, who accompanied Mr Nimmo on this occasion, and pointed out to him the disputed tract'.[24]

When Arrowsmith's *Map of Scotland,* incorporating Nimmo's corrections and revisions, was finally published on 25 June 1807, it cost £2,050 (not including the expense of tracing the Military Survey and several road plans that had been borne by the Commissioners). For his work on the whole map over two years, Arrowsmith received £300. It is a tribute to Nimmo's labour and specialist expertise that he received £150 for spending just five months working for the Commissioners. Arrowsmith's map became accepted as *the* authoritative standard map of Scotland for the next 50 years. Its importance was magnified by the extent to which it was used, directly or indirectly, by other cartographers, whether as a base map for geology, roads, or ecclesiastical boundaries, or for the ongoing work of the Commission for roads and bridges, which continued until 1862. All the Commission's progress reports to parliament carried a map by Arrowsmith, reduced from his 1807 map, showing roads completed or under contract in red, with those under consideration marked in green. For Nimmo, the results of his work had an even longer value in the form of the county boundary itself. In spite of its 'tidying' in the 1890s, and reorganisation in the 1970s through the local government acts, extensive parts of the southern boundary which Nimmo described still survive between the unitary authorities of Highland and Perth & Kinross. Nimmo would no doubt recognise its course and might indeed feel pleased that his perambulation, which put the boundary on the map, still has a value and purpose today.

Notes to Chapter IV

[1] I am grateful to Professor Noël Wilkins for his kind invitation to contribute to this publication, and to Dr Carolyn Anderson and Professor Charles W.J. Withers for their helpful comments on an earlier draft.

[2] In fact, no special county maps were ever published for the northern counties of Scotland, posing a particular problem for John Thompson and William Johnson in their Atlas of Scotland in the 1820s. See John Thompson, *The atlas of Scotland: containing maps of each county*, with introductory essays by Charles W.J. Withers, Christopher Fleet and Paula Williams (Edinburgh, 2008).

[3] Finally published six years after his *Map of Scotland, constructed from original materials*, National Archives of Scotland, RHP 14008 (London, 1807) as: Aaron Arrowsmith, *Map of the hills, rivers, canals, and principal roads, of England and Wales: upon a scale of six miles to an inch. Exhibiting most of the places whose situation has been ascertained by the stations and intersections of the trigonometrical survey* (London, 1813).

[4] Aaron Arrowsmith, *Memoir relative to the construction of the map of Scotland* (London, published by Aaron Arrowsmith 1807).

[5] John Watson, writing to his brother General David Watson in 1770, quoted in D.G. Moir (ed.), *The early maps of Scotland to 1850, Volume 1* (Edinburgh, 1973), 105.

[6] William Roy, 'An account of the measurement of a base on Hounslow Heath', *Philosophical Transactions of the Royal Society of London* 75 (1785), 385–478: 386.

[7] For further information and a facsimile of Roy's Military Survey see: William Roy, *The great map: the Military Survey of Scotland 1747–55*, with introductory essays by Yolande Hodson, Chris Tabraham and Charles Withers (Edinburgh, 2007, first published 1793).

[8] Alexander Nimmo, 'Historical statement of the erection and boundaries of the shires of Inverness, Ross, Cromarty, Sutherland and Caithness', in 'Third Report of the Commission for the Highland Roads and Bridges', Appendix U, British Parliamentary Papers (BPP) 1807 (100) Vol. III, 75–91.

[9] Alexander Nimmo, 'Journal along the North East and South of Inverness-shire. Ends at Fort William', f. 52–52 verso.

[10] Nimmo, 'Journal', f. 52 verso and f. 52.

[11] Arrowsmith, *Memoir*, 20.

[12] For more on Peter May see Ian H. Adams (ed.), *Papers on Peter May, land surveyor, 1749–1793* (Edinburgh, 1979). For information on the Annexed

Forfeited Estates see Virginia Wills (ed.), *Reports on the annexed estates, 1755–1769* (Edinburgh, 1973).

[13] R. Gibson, *The Scottish countryside : its changing face, 1700–2000* (Edinburgh, 2007), see especially chapters 13–16; M.L. Parry and T.R. Slater, *The making of the Scottish countryside* (London, 1980).

[14] J.B. Harley, 'The Society of Arts and the surveys of English counties, 1759–1809', *Journal of the Royal Society of Arts* 112 (1964), 43–6; 119–24; 269–75; 538–43.

[15] Christopher Fleet, 'James Stobie and his surveying of the Perthshire landscape', *History Scotland* 5 (4) (2005), 40–7.

[16] E.R. Cregeen, *Argyll estate instructions : Mull, Morvern, Tiree, 1771–1805* (Edinburgh, 1964).

[17] C.W.J. Withers, 'The social nature of map-making in the Scottish Enlightenment, *c.*1682–1832', *Imago Mundi* 52 (2002), 46–66.

[18] John Ainslie, *Scotland, drawn and engraved from a series of angles and astronomical observations* (Edinburgh, 1789). The earlier maps were of Edinburgh by John Ainslie and London by William Faden.

[19] C.J. Anderson, 'State imperatives: military mapping in Scotland, 1689–1770', *Scottish Geographical Magazine* 125 (2009), 4–24.

[20] Nimmo, 'Journal', f. 52 verso and f. 53.

[21] Cited in A.R.B. Haldane, *New ways through the glens: Highland road, bridge and canal makers of the early nineteenth century* (Newton Abbot, 1973), 59.

[22] J.B. Harley, 'The map and the development of the history of cartography' in J.B. Harley and D. Woodward (eds) *The history of cartography, Volume 1: cartography in prehistoric, ancient, and medieval Europe and the Mediterranean* (Chicago, 1987), 1–42:10.

[23] Nimmo, 'Journal', f. 31.

[24] Arrowsmith, *Memoir*, 20.

Chapter V

On Becoming an Engineer: Alexander Nimmo's Survey and His Engineering Career

~ Noël P. Wilkins ~

January 1811 was an important month for Alexander Nimmo. On its very first day he was admitted a Fellow of the Royal Society of Edinburgh, in whose roll his profession was recorded as 'engineer'. On the fifth of the month he was appointed an engineer on the staff of the Commission to Examine and Enquire into the Nature and Extent of the Several Bogs of Ireland, commonly called the Bogs Commission. His call to the Commission was probably the reason he was inscribed as an engineer in the RSE roll even while he was still Rector of Inverness Academy—and would remain so until June 1811. His appointment to the Commission signalled the end of his schoolmaster career; over the next 20 years he would become one of the most highly respected engineers in these islands.[1]

How could someone whose career was entirely that of a schoolmaster suddenly take on the mantle of an engineer? His prior practical engineering experience comprised mainly some experiments and observations he had made on the Moray Firth and in Loch Ness, together with the survey he completed of the boundaries of Inverness-shire, the topic of this volume. Ostensibly, the latter was a brief survey of land boundaries, not a full civil engineering apprenticeship as we know it today; he was not called upon to design or propose any roads or bridges (although he would do that), nor to construct any mechanical or other device. In fact, his claim to the title 'engineer' on the strength of this survey alone might seem slight in light of today's practice. But in those relatively early days of the industrial revolution it was not uncommon for enterprising and inventive persons to assume such titles in order to signify their skill in the mechanical or construction arts. In Nimmo's case, he was unquestionably a man highly educated, well connected in scientific and engineering circles in Edinburgh, and probably one of the few persons in Inverness who could have competently carried out the boundary survey to the requirements of the Commissioners for the Highland Roads and Bridges.

Nimmo was born in 1783, probably in Cupar Fife and lived with his family in Kirkcaldy. By contemporary accounts he was an unusually clever child with a deep and practical interest in mathematics. He studied for two years at St Andrews University (1797 and 1798) and subsequently transferred to Edinburgh University in 1799. The latter university was then experiencing a period of exceptional eminence in the pure and applied sciences. Among other notables, John Robison was Professor of Natural Philosophy (now called Physics), John Playfair was Professor of Mathematics, and Joseph

Black was Professor of Chemistry. James Hutton, a founding father of geology, had recently died in 1797, leaving a vibrant geological community in the city and the university.

Although there were then no formal engineering courses in any university, Professor Robison, whom some regard as the first true lecturer in engineering in Britain, was teaching and contributing articles to the *Encyclopaedia Britannica* (1797) on such varied engineering-oriented topics as carpentry, roof trusses, the compass, bridge arches and centring, dynamics, mechanics, steam and steam engines, strength of materials, rivers and water-works, and so on.[2] There is no doubt that the professors at Edinburgh attracted many students with aptitude in mathematics and the sciences who would subsequently become eminent and reputable scientists and engineers, men such as John Rennie Senior, his son George, Robert Stevenson, David Brewster, the young Irish engineer Richard Griffith and Alexander Nimmo.

Nimmo was well known to Playfair, who was probably the person who had recommended him for the bursary that had first enabled him to attend St Andrews University.[3] Nimmo attended Robison's as well as Playfair's lectures and remained in the university for a number of years. Later he was, according to a friend, '...one of those who were waiting in Edinburgh until Playfair should find situations in which they were likely to be useful both to the public and themselves.'[4] Up to then his formal education, however excellent, must have been one of 'book learning' alone. Eventually, Playfair obtained a teaching post for Nimmo in Fortrose

Academy, which he took up in 1802. Thereafter the two men continued in communication, so that when Nimmo wrote a scientific paper based on empirical observations of coastal processes in the Moray Firth, Playfair read it on his behalf to the Royal Society of Edinburgh.[5] That was Nimmo's first practical contribution to applied science that we know of. He broadened his research to include hydrographic investigations in Loch Ness which led him to propose the existence of significant deep-water counter-currents in the Loch.[6] These sparse records indicate, at very least, his keenness to gain practical experience and to apply scientific principles to the explanation of natural occurrences in his surroundings. But in truth, they are a very slim basis on which to found a practical engineering career. It is to his Inverness survey therefore that we must look for his first true engineering activity and for the genesis of his later engineering outlook.

Before considering the survey, some general observations are appropriate. Up to Nimmo's time and for some years afterwards, most engineers entered the profession from such practical activities as stone masonry, carpentry and milling, followed by an often long apprenticeship under an established engineer. Lacking such an apprenticeship, Nimmo had gained little familiarity with the drudgery of the long hours of work needed in constructing and perfecting real structures and had little practical, 'hands-on' skill. This limitation was to affect much of his later professional life and lead to the criticism that he tended to leave many of his enterprises unfinished. His friend, Charles Wye

Williams, would say of him 'All mind, and mind of a superior cast, it was almost natural that he should be indolent in following up the dull drudgery of carrying into execution the works and offspring even of his own genius'[7]—a kind way of acknowledging this failing. Nimmo seemed largely unaware of the consequence of leaving the completion of his schemes to men of lesser vision or competence. He, of course, might well have countered that his *modus operandi* was the same as that used by Telford and copied from him—he designed the overall schemes, leaving their practical implementation to resident engineers who would oversee their day-to-day construction and completion. Once, for example, in answer to a question about some harbours he was concerned with, Nimmo was driven to remark in frustration or exasperation (or perhaps embarrassment?) 'I think there is hardly a harbour that I have constructed, or been concerned in, that I have any doubt of the result, if they go on and complete the thing'.[8] As Wye Williams perceived him, probably correctly, Nimmo was a man of vision, ideas and theories, of plans and schemes rather than a mere journeyman engineer. These are aspects of his character that we might expect to find expressed, if only in embryo, in his Inverness survey.

On a broader front, the nature of the early period of the industrial revolution must be appreciated. As Birse points out, modern economic historians take the view that the industrial revolution was not the result solely of scientific discovery and technological innovation, but was 'the result of a very complicated interaction of political, agricultural, technological, economic and social factors, the relative importance of which is not fully understood.'[9] Ashton describes some of its agricultural aspects: lands were drained, enclosed and made arable; soil was being made more productive; new crops and crop rotation were introduced; tools were improved; transport to market was enhanced and regional specialisation in production and trade increased.[10] Elsewhere in this volume, Hunter has addressed the introduction of sheep to the Highlands, an important aspect of the industrial revolution in the region.[11] According to Ashton, although these advances did not, on their own, constitute an agricultural or agrarian revolution, they do indicate that the scientific character and social consequences of the industrial revolution applied as much to the agricultural sphere as to manufacturing. In consequence, anyone who applied scientific principles to agricultural improvement in its widest sense could just as properly be termed an engineer as one who applied them to the improvement of motive power or any other aspect of technological innovation. When we apprehend that Nimmo's first formative practical experience was his Inverness survey and his later surveys of Kerry and Connemara, it is not surprising that his subsequent engineering perspective would take on a pronounced social/agricultural/developmental aspect rather than a technological/mechanical one, although he would never entirely ignore the latter aspects. Nimmo became the engineer he did due as much to the nature of his experience in the places mentioned as to the prior academic formation he brought to his work. Nevertheless, he was one of the first modern engineers who entered the profession from an entirely academic rather than a practical background. That is why his Inverness survey is important in understanding his achievement.

Nimmo may have met Telford when the latter came to Inverness to initiate work on the Caledonian Canal. Indeed, according to Arrowsmith[12] it was Telford, the Commission's engineer, who recommended Nimmo for the boundary task, surely a significant endorsement. If true, it represents the first of many times in which Telford helped him and in which they would cooperate professionally. They would, for example, very soon afterwards in 1809 contribute a joint article on bridges to the *Edinburgh encyclopaedia.*[13]

As Rector, Nimmo was accustomed to taking the young Academy students 'into the field' to teach them practical surveying, much to the displeasure of the authorities, and something he would hardly have done willingly had he not the necessary surveying skills and interest.[14] Throughout the survey he was able to establish and delineate the boundary perfectly well using only the compass and some infrequent triangulation, even in the most difficult terrain. In this, of course, he was aided by the information he gathered from shepherds, tenants and factors as he progressed. He showed great confidence in their local knowledge and freely took their advice regarding local boundaries and marches. One detects no patronising of them in his journal, no criticism or disdain for their station. Much later in Ireland he would show, to an even greater extent, this empathy with the ordinary labouring people: 'I have sometimes slept in their cabins and had frequent intercourse with them...Really I am not inclined to think that the Irish are an indolent people...the value they set on their employment makes them industrious, sober and honest...' he told the House of Lords in 1824.[15]

The survey starts with some attention to the details of the boundary to the east of the shire. This area was certainly complicated (see p. xiv), with some farms straddling differing counties and often measured in arcane terms. But once into his stride as it were, Nimmo's progress became swifter and less fraught, permitting him to observe better the wider region and its agricultural and social organisation. He soon felt confident enough to criticise the engineering contribution of General Wade, the Irishman who had built over 240 miles of military roads in the Highlands in the early eighteenth century. Nimmo's criticism was that of a young man full of confidence in his own growing skill and ability, adjusting and adapting to his new surveying and engineering role. What that role was, and what Nimmo would make of it, are quite separate matters. The Commissioners needed an accurate delineation of the boundaries of Inverness-shire with the neighbouring counties to be entered on the map they had commissioned from Aaron Arrowsmith. Nothing else was demanded of Nimmo but to establish where the boundaries were and to indicate them accurately on the draft map, a task that Arrowsmith acknowledged he had performed 'with a zeal and intelligence surpassing the expectation of the Commissioners'.[16] While meeting these requirements, Nimmo recorded in his journal observations on agriculture, agronomy, industry, development, transhumance, sheep farming, town building, drainage, topography, cattle droving and so on in the Highlands, far beyond what was required or expected.

More than anything else, it was this broad perspective, this capacity to see beyond the immediate and the local, this interest in the wider scheme of things, which Nimmo would bring to Ireland in 1811 and which would mark all his activities there. This perspective fostered, in its train, a sense of hope and a spirit of enterprise that were largely lacking in Ireland at the time and that would come to be formally expressed later in the emergence of institutions like the Irish Ordnance Survey, the Hydrographic Survey and the Office of Public Works, in all of which Nimmo played an important role.

His remit from the Bogs Commissioners was admittedly much broader than his boundary task in Scotland. They required that their engineers not only examine the structure and extent of the bogs and propose means to cultivate them, but also ascertain the best lines of canals and roads to make the districts accessible for cultivation. All agricultural, topographical and physiographical features were to be laid down on suitable maps that the engineers were required to prepare. Starting in the Iveragh peninsula of Kerry in June 1811, Nimmo had completed by December the most comprehensive, thorough and positive account of the barony as it then was and had prepared what is generally recognised as the very best of all the Bogs Commission maps.[17] He followed this in 1812 with an account of the Barony of Dunkerron on the south side of the peninsula, together with another remarkable map, this one visually less dramatic but of greater delicacy.[18] Later the same year he would complete, to the same high standard, surveys and reports on the rest of the County of Kerry, but for these areas his maps would be more flat and monotonous, in keeping with the associated topography.

By any measure, the extent and comprehensiveness of his Irish surveys and reports exceeded anything that the Bogs Commissioners could have anticipated. As presaged in his Inverness survey, he possessed a keen eye for topography and natural resources, exhibiting an enterprising spirit for their development and an appreciation of existing agricultural and social practices, taking a perspective that located each area in a wide national context. He climbed mountains, sounded harbours, opened mines and criss-crossed bogs with tenacity and dedication. His Kerry reports encompassed 81 separate bogs comprising 150,000 English acres, proposed canals and railways to intersect the county, recommended river straightening, embankment, drainage and warping and commented on relevant social, agricultural, economic and political issues. In effect, he drew up an outline regional development plan where none was asked for, or even contemplated. It was a remarkable achievement by an erstwhile schoolmaster, after just 18 months in Ireland.

Throughout the Inverness journal Nimmo exhibits a strong interest in Gaelic etymology expressed in a series of marginal glosses, his description of words being both accurate and lucid, as we might expect from a schoolmaster. Arrowsmith certainly expressed his satisfaction with Nimmo in this matter, offering his own opinion 'that the preservation of the names of almost forgotten places ought to be reckoned a merit than a blemish in the map'—a sentiment with which many today would fervently agree.[19] In the journal and accompanying material, a mere 56 folios, Nimmo names over 300 places and topographical features of varying size and importance. He continued this linguistic and etymological interest into his Kerry reports but would mostly abandon the practice in his later account of Connemara. However, in all his reports, even to the very end, he retained a certain didactic, even schoolmasterish, tone that was evident in his social life also: Nimmo did not suffer fools gladly.[20]

On the Moor of Rannoch he proposed new roads to open up that impenetrable tract,[21] repeating the proposal in his covering letter to Rickman, even at the risk, he feared, of 'wandering from my proper subject'.[22] The irony of the implication that his road proposals could possibly be regarded as improper by Commissioners charged with 'making Roads and building Bridges in the Highlands' was probably not lost on either man. That Telford might resent this seeming intrusion into his own overall responsibility for proposing new roads was another matter entirely. When joined to the existing county road from Perth to Tummel Bridge, Nimmo's projected road to the west from Loch Rannoch to Kingshouse would open, in his own words, 'a complete communication across the centre of the Kingdom from the Eastern to the Western Sea'.[23] His second proposed road, the northern road from High Bridge near Loch Lochy to Killin in Perthshire would, he calculated, become the great drove road from the Hebrides and the north to the markets and fairs of Crieff and Falkirk in the south. He envisaged these roads as true *communications* opening the Highlands to the outer world, rather than mere *conveniences* joining one glen to another. Some years after his boundary survey the Commissioners would pay him £50 for marking out a route for his proposed High Bridge to Killin road.[24] Telford later claimed this road as his own idea, attributing to Nimmo only the estimation of its financial value to the drovers.[25] But Nimmo's proposal, recorded here in his journal and in his letter to Rickman, seems to contain the first mention of the road in writing, long before Telford's published claim.

Here, then, in the remotest, most inaccessible part of the kingdom, Nimmo was displaying a nascent taste for road building that would come to flower in the Irish countryside. In Iveragh in 1811 he would propose 28 miles of new or upgraded roads, including that around the foot of Drung Hill, which he later built and which is one of the most impressive sections of the famed 'Ring of Kerry' route. In the Barony of Dunkerron the following year he would propose over 37 miles of road and commence construction of the section between Blackwater Bridge and Derryquin. This, too, is part of the Ring of Kerry route. Elsewhere in the county he would propose a further 62 miles of new roads. His vision extended outwards to the Golden Vale of Munster (in Counties Limerick and Tipperary), which he proposed could be joined to the ocean at Killorglin by means of a canal (with intermittent inclined planes) from Kanturk in Co. Cork to Killarney and onwards to the sea by a new inland navigation to be constructed along the river Laune. All his roads and canals were conceived with a view to facilitating and enhancing the prevailing agricultural and mining activities of various localities and opening them up to the country at large. If his Rannoch road proposal was 'wandering from my proper subject' then he really wandered far astray in his proposals for Kerry!

It was in Connemara, Co. Galway in 1813 that the true breadth and comprehensiveness of his outlook, already apparent *in statu nascendi* in his journal, would be fully revealed. 57% of Connemara was mountain and upland pasture, 34% bog, 2% bare rock and only 7% arable. It was rightly considered one of the most uncultivated and uncultivable parts of Ireland; the Bogs Commissioners themselves held little hope that anything useful could ever be achieved there 'in its present state of desolation and abandonment' and they required only a general examination and brief report to be compiled. But Nimmo, who had traversed the Grampians with equanimity and crossed Rannoch Moor with optimism, was irrepressibly and resolutely confident facing this desolate tract. The sanguine outlook that he had attained on the very *Domum Britanniae*[26] had been finely honed in Kerry so that here in Connemara he would focus emphatically only on the advantages, as he saw them, of the region: the climate was mild; the mountains gave some shelter; limestone, shell-sand and seaweed were abundant; fuel supplies were inexhaustible; there were upwards of 20 safe and capacious harbours, about 25 navigable lakes of a mile or more in length, 400 miles of seashore and 50 of lakeshore. All that was needed was vision, expressed in a plan sufficiently comprehensive, to make the district productive and profitable.

Within 12 months Nimmo had thoroughly surveyed the district and written perhaps the most complete, objective account of it to date.[27] Few persons, before or since, shared the breadth of his vision, the liveliness of his enthusiasm or his boundless optimism for Connemara. He produced the first geological map of the area, one of the first regional geological maps in the United Kingdom.[28] But more significantly, he devised a thorough development plan for its infrastructure. This involved, for example, laying out 343 miles of new and improved roads, constructing new seaports and villages, devising new inland navigations, the opening of Lough Corrib to the sea by a canal and the extension of the road system outwards to the market towns of Westport, Ballinrobe and Galway. In Scotland, his Inverness survey had been only a minor element of Telford's great scheme for the Highlands, and Nimmo had only a small part in that scheme; here in Connemara the great scheme was entirely Nimmo's, and his alone: what Telford was to the Highlands of Scotland, Nimmo would strive to be to Connemara. By his reckoning the whole scheme would cost £42,000, a sum most unlikely ever to be made available. But he was not to know then, nor were any others, that fate acting through a local famine would bring him back to Connemara in 1822 to implement the scheme he conceived in 1813, which would become and remain to this day the foundation and template for the infrastructure of the region.

In 1818 Nimmo was elected a Member of the Royal Irish Academy and in the following year he, Telford and Robert Frazer, all Scotsmen involved with the Commission for the Highland Roads and Bridges, were the sole witnesses interviewed by the Select Committee on the State of Ireland as to Disease and the Labouring Poor.[29] There they recommended Scottish medicine (i.e. investment in public works in the manner of the Scottish Commissions) as a remedy for Irish ills. By coincidence, Charles Grant, previously MP for the boroughs of Fortrose and Inverness, son of Charles Grant Senior who was MP for the shire during Nimmo's Rectorship and his Inverness survey, was chief secretary to Ireland at the time. Understandably then, Grant was quickly won over by the Committee's recommendations, based as they were on the Scottish Highlands model he knew so well (his father had been one of the Roads and Bridges Commissioners) and on the testimony of persons he had known for a long time. The result was that the approach recommended by the Committee and endorsed by Grant would influence the nature and direction of public works in Ireland, and Nimmo's role in them, for the ensuing decade, culminating ultimately in the formation of the Board of Public Works in 1831. That the developmental history of Ireland in the first third of the nineteenth century owes so much to this nexus of Scotsmen and to the model of the Commission for Highland Roads and Bridges has been mostly overlooked. Other Scots engineers who also contributed in important ways were William Bald, who worked for Nimmo in Mayo (he also produced the Grand Jury map of Mayo and built the

Antrim coast road); Thomas Rhodes, who was taught by Nimmo (later becoming engineer for the Shannon navigation); Robert Stevenson, a classmate of Nimmo in Edinburgh and later a colleague (as well as a consultant on Irish harbours); the Rennies, father and son, and many more.

One outcome of the Select Committee was the setting up of the Commissioners for Irish Fisheries who engaged Nimmo in 1822 to survey the coast in order to identify sites suitable for fishery stations. With his by now predictable propensity to exceed the limited perspective and expectation of his employers, he used his new position to survey and draw detailed hydrographic charts of various bays, harbours and inlets around the entire coast.[30] This would bring him into conflict with the Admiralty which, with its own hydrographic survey at a standstill (the threat of a French invasion having disappeared by then), accused him of inaccuracy in his cartography. His coastal survey was immediately suspended. The complaint against him, however technically correct, may have been malicious, but it did not prevent him from completing most of his hydrographic charts and securing sufficient material for his *magnum opus*—his 'Map of St George's Channel and the coast of Ireland' with the accompanying book *New piloting directions for St George's Channel and the coast of Ireland*, published posthumously in 1832. This added to another remarkable survey and map he had completed earlier of the sea-bed sediments off the coasts of Ireland, England and France,[31] a major extension of his interests and influence outwards from Ireland.

In his public works in Connaught under the Famine Relief Act of 1822 Nimmo would exasperate the Irish Office through his characteristically liberal and wide interpretation of his duties. Under the Act he was appointed engineer of the western district (almost all of the province of Connaught) with instructions to implement famine relief measures of a minimalist nature. With his usual sanguine disposition he set about implementing the broad scheme he had conceived for Connemara in 1813, as well as making similar grand plans for the rest of Connaught. He stated:[32]

> One object I have been attending to…is to run main lines through the towns to the principal markets and seaports…we have one main object in view, to communicate between the extremity of the Royal Canal [on the river Shannon] and the seaports of Sligo, Ballina or Killala westwards, and the harbours of Connemara and Galway; another to communicate from the extremity of the Grand Canal [further south on the Shannon] on the lines which are now traversed by the mail coaches, to the towns of Castlebar and Westport, Galway and Gort.

And indeed, as if that were not enough:[33]

> I have prepared a plan…with a view, in conjunction with the operations in other quarters, to opening a great line of coach communication across the kingdom from Galway to Belfast.

It was an Irish equivalent of his earlier Scottish proposal to open '…a complete communication across the centre of the Kingdom from the Eastern to the Western

Sea.' Here again, in Connaught, we encounter an emphasis on *lines of great communication* rather than mere local *conveniences.*

The extent and scope of this broad vision appalled Chief Secretary Goulburn, who wrote to Lord Lieutenant Wellesley:[34]

> Mr. Nimmo certainly misunderstood the views of the Government in his original appointment. He embarked on works not so much calculated to afford employment to relieve immediate distress as to promote. . .more what appeared to him great national improvements. . .

Had Goulburn or Wellesley (or the Admiralty for that matter) read Nimmo's Inverness journal perhaps they would have realised that he was always likely to see the 'bigger picture' and almost inevitably would formulate his own plans on the widest possible scale. In this he was little different to Telford, who had been the first to endorse him for his engineering work. Telford had regarded his own efforts in Scotland as works of great national importance;[35] so too in this way did Nimmo see his works. Telford had been reprimanded for employing labour beyond what was approved and for expenditure that had not been budgeted; so too was Nimmo reprimanded. If it is excessive to say that Nimmo intentionally emulated his old mentor, at least his methods did not differ much from Telford's. Nimmo's perspective would never be narrowed by the pusillanimity of his masters, even less by that of Telford's 'official insects';[36] the breadth and grandeur of vision he had evinced and fostered on the Grampian eminences, on the awesome Moor of Rannoch and along the coast of Argyle were to stay with him throughout his professional life.

Nimmo began to disengage from Ireland around 1826, more because of his widening interests in England than any disillusionment with Ireland. One of his main engagements in Britain was the proposed development of new docks near Birkenhead. Along with Robert Stevenson, he proposed to join the Mersey estuary to the Dee estuary by means of a ship canal across the Wirral from Wallasey Pool to a new port to be developed at Helbré Isle. When first approached, Telford was off-hand about it, but once he saw the magnificence of what the others proposed he joined them readily.[37] The full proposal, signed by all three engineers, eventually came to nothing. Once again, it was Nimmo's expansionary propensity that had inflated a relatively minor dock project, originally envisaged by William Laird, into a magnificent scheme that appeared to pose a major threat to the Port of Liverpool. Liverpool Council responded by buying up all the land needed for the scheme and the speculation failed. Nevertheless it was, without doubt, a magnificent conception well worthy of Nimmo's all-encompassing vision and it could have been the crowning achievement of his engineering career. As it transpires, the last achievement of his career was to design the Dublin to Kingstown (Dun Laoghaire) railway, the first passenger railway in Ireland.

Nimmo's engineering career had started with his Inverness survey in 1806, involving both Telford and Rickman. It is a strange irony that the names of these three men would be associated again, 30 years later, in the debris of the Mersey/Dee debacle.[38] After the deaths of Telford and Nimmo, Rickman wrote of the affair in his introduction to Telford's posthumous autobiography.[39] In this he appears to accuse Nimmo, if only by innuendo, of deliberate duplicity towards Telford. That Telford felt betrayed and duped in the whole business is certain; that Stevenson was bemused and isolated in Edinburgh is equally true; but the innuendo against Nimmo was wrong and created one of the sorriest outcomes of the whole affair. Worse still, it seems to have caused a rift between Telford and Nimmo that may never have healed. Their long association, dating from the Inverness survey, seems to have ended peremptorily after the project collapsed.

Few persons will ever receive a eulogy like that written in the *Galway Advertiser* for Alexander Nimmo on his untimely death in 1832:

> His professional distinctions—Civil Engineer, Fellow of the Royal Irish Academy, Fellow of the Royal Society of Edinburgh, Honorary Member of the Geological Society of London, member of the Institution of Civil Engineers—were...but dust weighed in the balance when compared with the sterling talent and intrinsic merit of this excellent and lamented individual. Eulogium is unnecessary as the word *Ireland* alone will be both his most merited monument and suitable epithet. No man so well understood the remedies required for its practical evils, and the effects not alone of his foresight, but his actual works, will be felt long after the very remembrance of his name will have

> passed away. As a theorist and scientific member of his profession he has left *no equal* and in conclusion it may safely be said "the British Empire in general has sustained an almost irreparable loss".[40]

His physical contribution to Ireland included over 500 miles of road, more than 30 documented bridges and in excess of 53 piers and harbours, as well as his bogs and other surveys—a wonderful and lasting civil engineering legacy to his adopted country, in a career whose first shoots were so modestly nurtured in the Highlands of Scotland and recorded in his Inverness journal. It is altogether fitting and proper to publish his journal so that the remembrance of his name may not pass away in his native land.

Notes to Chapter V

[1] The first full account of Nimmo's life and works has recently been published: N.P. Wilkins, *Alexander Nimmo, master engineer, 1783–1832: public works and civil surveys* (Dublin, 2009).

[2] R.M. Birse, *Engineering at Edinburgh University, a short history 1673–1983* (Edinburgh, 1983).

[3] Wilkins, *Alexander Nimmo*, 2.

[4] R. Mudie, 'Notes by the conductor', in R. Mudie (ed.), *The surveyor, engineer and architect for the year 1842* (London, 1842), 2–6.

[5] J. Playfair, in 'Minutes of the Royal Society of Edinburgh', 3 December 1804, National Library of Scotland, Acc 10000/4.

[6] Alexander Nimmo, 'On the application of the science of geology to the purposes of practical navigation', *Transactions of the Royal Irish Academy* XIV (1825), 39–50: 42.

[7] Anonymous [C. Wye Williams] Obituary of Alexander Nimmo. *Dublin Evening Post*, 28 January 1832. (Original manuscript in University College London, Greenough Papers)

[8] Alexander Nimmo, 'Evidence', in 'Minutes of evidence before the Committee

on the Norwich and Lowestoft Navigation Bill', British Parliamentary Papers (BPP) 1826 (396) Vol. IV, microfiche (mf) 28.24–6, 407.

[9] Birse, *Engineering*, 42.

[10] T.S. Ashton, *The Industrial Revolution 1760–1830* (revised edn, London, 1962).

[11] James Hunter, 'The Scottish Highlands and Ireland in the time of Alexander Nimmo' in this volume, Chapter III.

[12] Aaron Arrowsmith, *Memoir relative to the construction of the map of Scotland*, National Archives of Scotland GD9/40/1 (Edinburgh, 1807).

[13] Alexander Nimmo and T. Telford, 'Bridge' in D. Brewster (ed.), *The Edinburgh encyclopaedia* (Edinburgh, 1811).

[14] Robert Preece, 'Alexander Nimmo: Rector of Inverness Academy' in this volume, Chapter II.

[15] Alexander Nimmo, 'Evidence', in 'Minutes of evidence taken before the Select Committee of the House of Lords to examine into the nature and extent of the disturbances which have prevailed in those districts of Ireland which are now subject to the provisions of the Insurrection Act and to report to the House', BPP 1825 (200) Vol. VII. mf 27.60–63, 179. Hereafter 'House of Lords Report'.

[16] Arrowsmith, *Memoir.*

[17] Alexander Nimmo, 'The Report of Mr. Alexander Nimmo on bogs in the Barony of Iveragh in the County of Roscommon', Appendix 5 in 'Fourth Report of the Commissioners for the Bogs of Ireland', BPP 1813/14 (131) Vol. VI, Pt 2, mf 15.33–6. Hereafter 'Fourth Bogs Report'. (In its title, the published report mistakenly locates Iveragh in Co. Roscommon.)

[18] Alexander Nimmo,'The Report of Mr. Alexander Nimmo on the remaining bogs in various parts of the Counties of Kerry and Cork', Appendix 6 in 'Fourth Bogs Report'.

[19] Arrowsmith, *Memoir.*

[20] Anonymous [C. Wye Williams] Obituary of Alexander Nimmo.

[21] Alexander Nimmo, 'Journal', f. 37, in this volume,

[22] Alexander Nimmo, signed MS letter to John Rickman, dated 12 Oct 1806 in 'Scotch Boundary's', f. 53 verso in this volume.

[23] Nimmo, 'Journal', f. 37, in this volume.

[24] 'Fifth Report of the Commissioners for the Highland Roads and Bridges', Appendix U, BPP 1810/1811 (112) Vol. IV, mf 12.24.

[25] Thomas Telford, 'Report and estimates relative to a proposed road in Scotland from Kyle-rhea in Inverness-shire to Killin in Perthshire by Rannoch Moor', (London, 1810).

[26] Nimmo, 'Journal', f. 30.
[27] Alexander Nimmo, 'The report of Mr. Alexander Nimmo on the bogs in that part of the County of Galway to the west of Lough Corrib', Appendix 12 in 'Fourth Bogs Report'.
[28] Alexander Nimmo, 'Geological Map of Connemara, Ireland'. Notes captioned by G.B. Greenough 'Nimmo', 'Connemara'. Manuscript map archived in the Geological Society of London, LDGSL 999 (London, ND). A coloured reproduction of this map is given in Wilkins, *Alexander Nimmo*.
[29] 'Second Report from the Select Committee on the State of Disease and Condition of the Labouring Poor in Ireland', BPP 1819 (314) Vol. VIII, mf 20.67.
[30] Wilkins, *Alexander Nimmo*, Chapter 9.
[31] Alexander Nimmo, 'On the application of the science of geology', 42. A coloured reproduction of this map is given in Wilkins, *Alexander Nimmo.*
[32] Nimmo, 'Evidence', in 'House of Lords Report', 171.
[33] Alexander Nimmo, 'Report on the progress of the public works in the western district of Ireland in the year 1829', in 'Public Works Ireland', BPP 1830 (199) Vol. XXVII, mf 32.194, 3–4.
[34] H. Goulburn to Lord Wellesley, 23 May 1823, British Library, Wellesley Papers, Add 37301 f 90.
[35] A. Gibb, *The story of Telford* (London, 1935), 60–88.
[36] 'Official insects' was an expression Telford used for bureaucratic officials, made in a poem he wrote on the death of the poet Robert Burns.
[37] A full account of the plan and its ramifications is given in Wilkins, *Alexander Nimmo*, Chapter 13.
[38] The Mayor had perceived the the proposed ship canal across the Wirral from Wallasey to the Dee estuary as a threat to the commercial importance of Liverpool docks and the Mersey, since it could potentially transfer port fees and dues to the port of Chester, which controlled the Dee estuary. To prevent the scheme going ahead, Liverpool Council had to spend almost £158,000 and Tobin, Laird and others who had bought the land on speculation made large profits from its sale. George Nimmo—Alexander's brother—was alleged to have made a surveying mistake that doomed the project so that the engineers lost a magnificent opportunity. In a subsequent Court of Enquiry, Liverpool Council was accused of being duped in the whole affair. So, in the end, few came out unscathed from it.
[39] J. Rickman (ed.), *Life of Thomas Telford, civil engineer, written by himself* (London, 1838).
[40] Anonymous, Obituary of Alexander Nimmo, *Galway Advertiser*, Vol. XIV (5), 28 January 28 1832.

JOURNAL ALONG THE NORTH EAST & SOUTH OF INVERNESS-SHIRE.

ENDS AT FORT WILLIAM

by

ALEXANDER NIMMO

{f. 2}

In the Spring 1806 I had the honour of being appointed by the Commissioners for making Roads and building Bridges in the Highlands of Scotland,[1] to make a survey of the limits of the 5 Northern Counties and to lay down the outline on the map now publishing under their authority.[2] Having accordingly made a journey through these districts in the summer following I shall now proceed to give a brief but faithful detail of my progress and the various data from which my remarks or corrections were derived.

In doing this I shall chiefly copy the notes I occasionally took in my pocket book—which were not at first drawn up with any methodical or even connected arrangement—and having already drawn up a concise historical account of the successive division of the district in question, where I have also described the boundary of every county in order,[3] the present journal will exhibit the greater part of the materials from whence that account especially of boundary has been formed.

As the inconveniences to which I was sometimes subjected can be of no consequence to the public, I shall not therefore touch upon them unless where they appear necessary to elucidate my subject.

{f. 3}

May 1st

At the 20th April meeting of the County of Inverness I took the liberty of writing the convener expressing my sentiments concerning the method which might be adopted in communicating to the person appointed to survey that information which he might require and being sensible that there were in some quarters undivided commons of considerable extent between Counties, that the respective counties might appoint some individuals, as commissioners, to point out the extent of the commons or to agree upon some line to be considered as the county boundary until the matter was adjusted by a legal division, that in such a delicate case the surveyor might not be left entirely to his own judgment.

The meeting, however, did not consider it competent for them to interfere with the limits of private property by which the boundary of the counties is generally determined and satisfied themselves with writing a circular letter to the border heritors requesting them to take such methods of giving the requisite information to the surveyor as might be necessary.

To some of these letters, answers were returned, but from the greater part of the boundary, the convener had no reply. This was peculiarly felt in the case of the Duke of Gordon whose property limits the county for a very considerable extent.

{f. 4}

It was however a fortunate circumstance that the limits of particular properties were also those of the county and were in most cases distinctly marked out by land marks perfectly well known to the farmers, shepherds etc on the spot. Information obtained from them of the boundary etc in their immediate neighbourhood might therefore be relied on as perfectly legitimate although their knowledge with respect to more distant quarters might not be very precise.

I take notice of this circumstance lest it should be said that my description of the boundary is in any instance incorrect and should it be found so—though I have little reason to believe it will—I cannot but think myself exonerated from blame in being left in most cases to hunt for the necessary information amidst the difficulties of a desert and unpeopled border district with the few inhabitants of which I was generally obliged to converse by the medium of an interpreter.

{f. 5}

May 20th

Receive from London the skeleton map with instructions from Mr. Rickman.[4] I did not at first think it practicable to perform more than the perambulation of the borders of Ross, Cromarty, Caithness and Sutherland in the short time I was able to spare for this purpose. Various circumstances seem now to make it necessary that I should also examine the East and South boundary of Inverness Shire and even begin my journey in that direction.

Sir George McKenzie[5] having written repeatedly for a copy of the skeleton map, so far at least as relates to Ross Shire, I have taken it to Coul and shown it to him. He proposes measuring a base in the Black Isle or even on the top of the mountain Ben Wyvis with a view to its verification. It appears to me impossible to accomplish this with any degree of accuracy in the short time that I can bestow upon such an object.

The map seems amazingly correct even in the more minute parts, and in its general construction. Much more may be learned from a few astronomical observations than from a trigonometrical operation, unless carried on upon a scale the magnitude of which my present arrangements entirely preclude me from engaging in.

{f. 6}

In compliance with the proposal of Simon Frazer Esq. of Farraline[6] I have shown the map to the Honorable Col. Frazer of Lovat[7] and him. Lovat mentions that there are extensive plans in his possession of his estates in Glen Strathvarer etc which bound with the County of Ross and which he proposes I should examine as likely to give me more precise information than it was possible to obtain by actual perambulation.

Saturday May 21st

Agreeable to appointment I waited on Lovat at Beaufort Castle, who very politely showed me the plan of the barony of Glenstrathvarar, taken during the annexation of this Estate in 1758[6?] by authority of the Board of Exchequer, by Peter May.[8]

This plan seems executed with great care, is about 12 feet by 4½, scale 10 Scots chains per in., or about 7 inches to a mile.

It comprehends Loch Monar and the course of the river as far as Culigran, improperly named on our map Calgran.

The boundary to the west and north with Seaforth and to the north-east with Fairburn, forms the boundary of Inverness Shire with Ross. A sketch of this boundary I marked on the skeleton map.

The comparison of this survey with the map affords a proof of the general accuracy of the latter. There are however a few corrections necessary. The river which runs northwards into the middle of Loch Monar must be erased. The branch which falls in near Lochil or rather Inch Lochel must be drawn

{f. 6 verso}

Sgur na Lappich. Sgur—a sharp peaked pyramidal mountain. Lappich—spotted or streaked?

This mountain is very high—say 3800 to 4000 feet above the level of the sea.[9] The name applies only to the northern face with propriety, the southern being named —— [blank][10]
It abounds in crystals of quartz. Many patches of snow continue perpetually on the north side
Craig mor tuil n' Lochin—Great rock by the pools, Tuiln?
Awin—a river. Measgach—drunken.
Meal—a mass or heap. Bui—yellow.
Bein—a mountain of considerable magnitude, not peaked.
Muich—a hog.
Corra Glas—Gray Corry. A large ravine or rather circular valley. Cor—a kettle.
Rui—a shealing or summer hut—a sequestered district where cattle are pastured in summer. The word enters often into composition.

{f. 7}

to the west, and towards the source it turns away to the South (rising from the north side of Sgur na Lappich Mt). Any other corrections are of little moment: a few streamlets are drawn too far; I have marked a cross where they terminate. Some are omitted: I have inserted them.

Boundary

The boundary commences with a hill to the west of the summit of Scor na Lappich (called Craigmore Tuiln Lochin) which is the boundary of Lovat and Chisholm (Inv.) with Seaforth (Ross), proceeds NNE (crossing a road which leads, by the side of Awin Mish Kelsey[?], into Kintail) to the summits of a little hill called Meal n' bui, to a rivulet named Ault na Crilie and so to Loch Monar and down the Varar.

The river Varar forms the boundary only for a little way before we arrive at Inch Lochel. The outline proceeds to the north by a rock named Meal Tarnach (Bulls Hill), along its summit E to a steep rock named Bein Muich whence by the summits of the following mountains: NE—Scur a Muich; SE—Scur a Corra Glas & Scur na Rui; NE—Scur na Corra Charabie to the mouth of the rivulet of Alt na Corra Charabie, which falls to the Orrin. Lord Seaforth etc claim [*sic*] half a mile farther down. Proceeding down the Orrin about a mile we have to the S, the hill of Carn a Cossich by the summit of which SE to Meal Goram Loch, S to Scur a Lowlan [?] and to Carn a Cavelach—and so to the eastward by the boundary between Chisholm and Fairburn about a mile and a half North of the river at Culigran where this survey ends.

N.B. Place Culigran and most of the places in similar narrow glens, immediately on the bank of the river. Several places are marked on the map and surveys which by the introduction of sheep are now uninhabited.

{f. 7 verso}

Glas—Grey.
Leitter—a woody bank.
Mam suil. Mam or Maum is usually applied to those high and elevated paths which run across the summits that divide two parallel glens.
Suil—the eye, a prospect.
This mountain is generally considered as one of the highest in all that neighbourhood. The famous Loch 'n uain or Green lake is in the bosom of this extensive mountain. That the lake always has ice in it Mr. F. asserts as a fact consistent with his own knowledge.
Beinn Mhian—Middle mountain.

{f. 8}

So far the information to be derived from this plan. There is an accurate survey of the parish of Kilmorack sworn to by Provost Brown—pupil and nephew to the aforesaid Peter May—which exhibits the boundaries towards Beauly—now in the hands of Rob[t] Dundas Esq. W.S.[11] Edin[b] sent there by Lovat who had it from the exchequer, on account I believe of a process now depending, about the very boundary in question.

Saturday

Mr. Falconers—Farmer, Drumriach near Kirkhill.

Mr. F. has long resided in the head of Strath Glas. Loch Mealardich (Lake of round high ground) is the proper name of what is called Loch Mayley in our, Loch Maddy in Ainslie's map. Loch Mayley is a small lake in Glen Strath Varar at a place named Muille.

Chisholm's property, called Glas letir is in Ross Shire having been bought from some branch of the family of McKenzie. The boundary of Inverness runs south from Scur 'n Lappich to the mouth of Loch Glasletir whence proceeding by the lake to a small rivulet that falls into its head on the South, it ascends to the summits immediately west of Mam Suil Mt. It proceeds by these summits round the head of Glen Grivie.

For Loch Benevach insert L. Benevian. Affrick is usually spelled Affarig.

{f. 9}

Monday June 2nd—Inverness

Hired a young man from Urquhart, named Duncan Grant, well acquainted with Gaelic, to accompany me.

Having now arranged matters for my journey and the route I was to pursue I wrote to the convenors of the several Counties of Nairn, Moray, Banff, Aberdeen, Perth and Argyle, stating the time I proposed setting out viz. Friday the 6th ins^{t} to begin at Ardersier, near Fort George, and from thence to proceed southward along the outline of the county of Inverness—mentioning also to each the period at which I expected to be upon the limits of his County.

As I could give no precise information to the counties farther north, I forbore writing them until I should be further advanced on the Survey. This I had afterwards reason to regret especially in the case of Ross in discovering the northern boundary of which I had at first some difficulty although at length this was completely overcome.

Tuesday

In the meantime having heard that the plans of the forfeited estate of Cromarty were in the hands of

Mr. Donald Cameron Writer in Dingwall[12] I rode thither to examine them.

By the way I stopt [*sic*] at Ferrintosh, an insulated district in Nairn Shire, where by the assistance of Mr. Kinloch factor for the proprietor, and a

{f. 9 verso}

Bein Uais—wondrous mountain.

{f. 10}

survey of the estate which is now nearly executed, I was enabled to lay down that district on the map with sufficient accuracy. I had previously seen the plans of the estate on the South, and the limits being everywhere distinctly ascertained, no difficulty can arise in that quarter.

This district may contain about———[blank] square miles of which the part towards the Connan is extremely populous and comparatively well cultivated. The upper part was till very lately an undivided common moor. It for many years enjoyed the privilege of distilling its barley duty free, a right now purchased by Government, which, while it lasted, greatly promoted the prosperity of Ferrintosh.

Wednesday—Dingwall

The only plan in the hands of Mr. Cameron I found to be the grazings of Garbet and Glenscaingh including the greater part of the Mountain Ben Wyvis in extent about 6 miles by 3. From this plan I marked out a sketch of the boundary on the map of that side of the Lordship of Castle Leod, or Strathpeffer, which forms part of the County of Cromarty. No information could be had concerning the remaining plans. Mr. Cameron pointed out to me however, the greater part of the eastern boundary of that estate. The former mentioned particularly the gentlemen appointed to communicate the necessary information in every part of the county. The latter inclosed an extract of Ainslie's Moray,[13] with a few observations and corrections of names, referring me chiefly to the factor for Sir James Grant in Strathspey, who by his local knowledge of the country, was indeed well qualified to give me any necessary information.

{f. 11}

Thursday

Meantime I went to Culloden and was shown by Mr. Kinloch factor for Forbes of Culloden, some plans of his farms on the water of Nairn, several of which are situated in the County of Nairn.

I received the telescope by the carrier from Aberdeen whither it had come by the mail. This I afterwards found one of the most valuable instruments I could have in my possession. The sextant and horizon had already arrived by sea. From their bulk they became very inconvenient in travelling. The trouble of unlashing them when fixed on horseback prevented me from taking several observations that might have been of service in constructing or correcting the map.

Fortunately in marking the outline I seldom found them necessary. The number and accuracy of the physical positions on the map enabled me in many cases to speak with more precision than even many of my guides when we were not on the spot.

I took along with me a small pocket compass. It was not fitted with sights, but was otherwise of the greatest service in so much that at length in drawing a slight sketch of any little district and in recognizing objects when on elevated stations, I came at length to rely chiefly on that and the telescope.

{f. 11 verso}

(SEE P. XIV AND XV)

Nimmo's sketch of the boundary to the east of Inverness captioned by Nimmo: 'A sketch map in pen of the boundary from the Moray firth to Budzeat.'

{f. 12}

Friday, June 6th, 1806

Left Inverness for Ardersier, near Fort George. Arrived there by 12, found Mr. Falconer of Nairnside from the County of Nairn, and Mr. Macpherson of Ardersier, from the Counties of Nairn & Inverness expecting me. The day very boisterous. In the evening we rode out and examined the boundary from the Ft. George Road to Nairn, southward to the Loch of Flemington, including the district called Calders Brackla.

Saturday

Next morning we proceeded to Kilravoc Castle where Mrs. Rose of Kilravoc showed me a plan of the mains of Kilravoc which give the boundary by Croy to the river Nairn and down the S. side of the bridge. Rode along that part of the line with Mr. F. The estate of Kilravoc (in Nairn) includes several little patches on the S. side of the water towards Budgate or Budzeat.

At Calder Castle examine a plan of the estate of Lord Cawdor from which we have the boundary, 1st from the sea to Kilravoc's property, 2nd round Bracklich, 3rd round Budzeat or Budgate, 4th round the Thain's moor and Leonach.

From this plan I was enabled, with the assistance of several angles and bearings taken from the tower of Kilravoc Castle to draw a more correct sketch of the country between Fort George, Nairn and Calder Castle, the map and plan of the canal being very defective in that quarter.

Return to Kilravoc, by a part of the Budzeat line, proceed to Cantray. No information can as yet be provided respecting Rose of Holms property in which the boundary line is somewhat intricate.

{f. 12 verso}

Sketch map No. 2. An incomplete sketch, in pencil, of the area around Cantray.

{f. 13}

Monday—Tuesday, June 9th, 10th—Cantray

The information I have been hitherto able to procure respecting the boundaries in this quarter is extremely imperfect. The farms seem alternately in Inverness and Nairn, but without any kind of order or connection. Thus on the south of the river, we have not only Budzeat but several farms above it in Inverness, as well as Cantray and Holm on the north side. Now we have Cantradown and several farms above it on the south, and the single farm of Dalgramich on the north side in Nairn Shire. Kinrea is in Inv^{ss}, Cantra Bruich in Nairn etc.

I have examined in company with Mr. Alex Grant, factor for Cantray, the plans of several parts of this estate, but comparatively with little success. Mr. Grant was only lately appointed to his present situation, and his local knowledge is not very precise. We expect much assistance from the late factor, Capt. McPherson Culdoich, but who is not now at home.

We have at length stumbled upon a clue to this labyrinth, it is a claim of enrollment as a freeholder of Nairn by the late Dav[d] Davidson Esq. of Cantray.

This document founds the claim upon the barony of Clava *olim* Clavalg, which is 'retoured' in the cess books of Nairn, and comprehends the farms of Clava, Dalroy, Cruagorstan and Cantradown on the south side of the river, with Dalgramich , Cantrabruich, Little Cantra, E and W Urquhil, or Uruchil on the north side.

The instrument goes on to state how much of this is in the claimants [*sic*] possession.

It became now only necessary to determine the boundary of these farms and this was done with comparative facility. A great part of the

{f. 14}

outline lay in our view, and could likewise be had from the plans. Some difficulty arose in determining the boundary between Dalroy and Culdoich on the west of the Clava estate, from the ignorant apprehensions of some country people to whom we applied for information, but this was soon removed by Captain McPherson, who met us with an intelligent guide and who pointed out to me the western outline of Clava and Urquhil. The boundary in the hill country to the south is not however, very distinct.

The southern boundary towards the muir of Culloden is a little uncertain, there being some common pasture for all Forbes of Culloden's tenants where no outline is marked. This is the case behind Cantrabruich, Little Cantra and the Urquhils in particular.

Wednesday—Inverness

As the cess books of the several counties seem likely to give valuable information, I have procured a copy of the valuations of Inverness, Ross etc. This I have examined with some care, but as the properties are generally valued 'in cumulo' with reference only to the name of the proprietor I have derived little benefit from them with respect to boundaries.

I have also consulted the rolls for the assessed taxes in the hands of Ja^{s} Grant Esq Provost of Inverness[14] and collector of taxes for the county, but with little success.

A third book which I have examined, but with little advantage, is that called

{f. 15}

the book of Splittings. In cases of the partial sale or transfer of property, a committee of commissioners of supply are [*sic*] called to split or divide the valuation affixed on the whole estate into parts proportioned to the conceived value of each lot. It is chiefly in this way that the valuation of single farms may be discovered. A record is kept of their proceedings titled the book of Splittings, and from this I expected to be enabled to throw much light on the division of the Counties of Inverness and Ross, which took place little more than a century ago. Unfortunately however, most of the records of the Sheriff clerk's office in Inverness were destroyed by fire about a century ago. Since then the record of splittings, which has no index etc, seems hardly worthy of consultation for the above mentioned purposes.

I now found a letter waiting me from the clerk of supply of Perthshire—tis couched in terms rather singular—and considering the notice I have given of entering on this survey as too short, he concludes by protesting against such proceedings.

I had given them notice nearly three weeks before I was on the border of Perthshire. I forebore giving any answer to this letter. Indeed I could not well give any. Meantime I again prepared for a longer journey than before, and as the ensuing eclipse of the sun[15] might perhaps be turned to good account, I increased my baggage with the Nautical Almanack[16]and Mackay's Treatise on the Longitude.[17]

{f. 15 verso}

Bariven—the tomb of Ivan or Ewan, a Danish prince said by tradition to have been killed besieging the fort on the neighbouring hill called Dun or Dunevan.

Dun—a little fort. Hence Ach 'n Dun, field of the fort. The enceinte is still visible about 63 yards by 30. Not vitrified.

{f. 16}

Thursday June 12th

Left Inverness for Auchindown, Dulsie and Grantown. I intended to pursue this route by Strathspey to Perthshire thence to Ballachulish in Argyle, to Fortwilliam and Loch na Gaul, where I intended to take boat for Kintail and return by Strathglass to Inverness.

Cross the Nairn at Kilravoc Bridge. The old military road winds up the hill by several traverses and is very rough. General Wade[18] was no engineer, although what he did was no doubt much for the benefit of the North. This road is now quite abandoned for the county road which runs on the east of Calder Castle and falls on the old line near the bridge at Highland Boath, 2 miles before crossing Findhorn. Soon arrive at Auchindown. Major Grant and family at church, being sacrament at Calder Kirk.

The old church of Bariven about a mile to the south-west. It has been abandoned for nearly a century, the chapel of Cawdor Castle being now used as the parish church. Day rainy and boisterous. The neighbouring rivulet is called Ault Derg, the

Red Burn, and an hour or two convinced us that the name was sufficiently appropriate.

Friday June 19th

Went to the hill called the Dun, commanding a very extensive prospect, where Major Grant assisted by a tenant of the name of William Rose born in the farms to the south of Ault Derg, pointed out the line of boundary. It must be observed that as most of these farms pasture in common, the line of boundary is altogether uncertain when we pass the arable land. In such cases we adopted that which appeared most natural, viz.

{f. 16 verso}

Auchindown Friday 13th June 1806

XII	29' 50"	D. Alt 110	51	50
	35 10		35	40
	37 55		24	40

Index Error 21' 40

XII	43 40	110 6 11
	45 56	109 55 15

Frequent clouds.
The watch to be corrected.

{f. 17}

to produce those of the arable lands, to the limit of the common pasturage. I took an observation of the sun this day, which may suffice to determine the latitude of Auchindown. The watch I used showed seconds but not being adjusted, it can only be depended on for the elapsed time.

A disagreeable accident happened after leaving this place. When leading our horses, the young man who accompanied me as a servant happened to let his go and did not recover her for some miles. When I afterwards examined the instruments I found the brass cap slipped from the bottle of mercury and a good deal spilt in the case. The subsequent jolting converted a part of it into a black oxide which did not however prevent me from having a horizon tolerably clear.

N. Prolong the county road by Calder to near the bridge at Highland Boath—say ½ mile from it—remove this water (called the black or muckle burn and not the water of Brodie) to within 2 miles of Findhorn. The Fort George and Edin[b] road leaves the Grantown at Ballinault Bridge and runs SE p. comp.[19]

Enter Moray at Etrich mor and Inverness at the bridge of Dava. There is no road from this place to Duthel as on the map.

Arrive in the evening at Grantown, a thriving village, upon Sir James Grant's estate in Strathspey.

{f. 18}

Saturday, June 14th Grantown

Breakfast with Mr. Ja[s] Grant at Heathfield. Mr. G. has been for many years factor on the estate of Strathspey, and has been present at settling most of the 'marches' or boundaries of the estate.

Indeed these settlements, or striking of marches, are in many cases of comparatively late date. The value of the pasture land was formerly little understood in most parts of the highlands. The tenants kept as many cattle as they could winter on the fodder and of their arable land. In the spring they were fed on the neighbouring pastures and as summer approached, they removed with them to the distant and sequestered glens called Arrie or Aosach where they dwelt in small temporary huts called Bothies or Shealings, living chiefly on milk till the approach of Autumn warned them to return to the lower vallies [*sic*] or straths where their little [*sic*] corn was now ready for the sickle and a sufficient stock of herbage for their cattle during the latter part of the year. Indeed this is still the practice in a greater or less degree in all the districts where the sheep system has not been adopted. For such a vague method of culture as this there was little need for accurate boundaries. The most convenient access was the chief bond of connection between districts, while an impassable mountain, or an unfordable river, seemed the only boundary of property.

{f. 19}

Of late however, lands have been rented with a view to the pasture alone and it became necessary for the proprietors to have their marches more distinctly settled. Boundary titles were unknown, but evidences were adduced to show where the tenants of particular districts were in the habit of 'shealing' and the property decided accordingly. Many instances of this have occurred of late in Scotland. Difficult cases are commonly settled by arbitration which in the Scots law is final.

Mr. Grant being unable to accompany me himself procured the Baron Officer or ground bailiff, William Grant, for this purpose, with whom I traversed this day the greater part of the western boundary of this insulated district of Inverness Shire. We had a view of a considerable part and of the meeting of Nairn, Inverness and Moray from a mountain on the east side of the Lake called Loch 'n Dorb. From thence we descended to Dava and proceeded by the hill of Attindow and after crossing with difficulty some extensive bogs, we returned by the hills to Grantown. Having described this boundary in another place[20] I shall not repeat it here.

Monday June 16th—Castle Grant

I went with Mr. Grant to the Castle to examine several plans of different parts of the estate. Few of them extended beyond the arable lands.

~

Castle Grant June 16th 1806

XII 11' 50"   III 58' 10"
XII 15   III 39' 40"
Error 21' 15"

Bynac Mt the boundary with Banff is on the meridian
XII 20' Sun on Carn Goram
XII 22' 15"   III 26' 15"
23 15   20 35
30 19   3 5
Error 21' 15" by hills.

Correct the horizon glass.

Solar diam. Fore	51' 41"	
Back		47' 30"
		64 10
	Diam	32 5
	Error	19 35
Do. By hills		20 30
		40 5
	Mean	20 2

{f. 20}

The annexed observation will give the latitude of Castle Grant. It was taken in a field about 100 yards south of it. I had afterwards made some preparation for observing the eclipse of the sun here, but was unfortunately disappointed, the latter part of the day being quite overcast.

Tuesday

By the help of Mr. Grant I had now determined the boundary to the north of this district from the Spey at Culcuich to the Bridge of Bruin. We found the outline marked with sufficient distinctness on the map. Some alteration was necessary in the outline returning from this bridge by Connage to the Spey, which is the north boundary of the Lordship of Abernethy. I this day perambulated that outline with William Grant. ——-[erased].[21]

On the south of Abernethy there is no definite boundary excepting between the arable farms of Rothymoor (Moray) and Gartenmor (Inv.). Mr. Grant proposes therefore to draw a strait [*sic*] line from that march to the Mt Dagram in Strathdown or Glen Awin.

Again the districts called Caiploch and part of Ailneck seem to belong with most propriety to Strathspey, having always been a part of the pasturage of the farms there. On this account we abandon the summits of the hill of Abernethy and crossing the Ailneg at a place called Ca na Churachar directly to the eastern shoulder of Dagram Mt we then follow the summit level to Bynac Mt, and proceed by the march stones of the Duke of Gordon and Sir Ja^{s} Grant, to Cairn Goram Mt.

{f. 20 verso}

Davoch—from Daimh, oxen and ach, a field. A portion of land which required 4 team of oxen to plough it in a season—usually about 100 acres. The word frequently enters into composition, as Culdoich, Davochmaluac, Dochgarroch, Dochfour etc.

{f. 21}

We next investigated the outline of the Parish of Duthel situated in the valley of Dulnain, and which belongs wholly to the County of Moray. The outline from the South of Nairnshire to the Spey, near the mouth of Dulnain, I made the subject of a new perambulation during which we met with the following singularity. In the farm of Laggan, the boundary is perfectly indeterminate. All that can be said is that two thirds of it are in Moray. The farm was originally cultivated in runridge, by the tenants of Ballintomb and Gaich, of which the former is in Moray, the latter to the south of it is in Inverness. It was supposed to consist of three 'Aughten parts' that is three eights of a Davoch. Two of these were considered as belonging to Ballintomb, the third to Gaich, but as this arable land lay in scattered ridges or patches, a new division would probably be made every season and thus the boundary remains to this day undetermined. Tis said that the house is in Moray and the barns in Inverness.

One would be tempted to think this a rude contrivance for connecting the two parts of Inverness Shire without disjoining those of Moray, for this farm with some common pasturage extends to the bridge of Curr on the Dulnain, as do those of Finlarig in Duthil Parish and Ballintomb at the mouth of the river, which are in Moray, while beyond the bridge and occupying the angle between Dulnain and Spey, are the farms of Curr, Clury and Tullochgorm which form part of Inverness Shire and the last of which completes the connection to Gartenmor on the opposite side of the Spey.

{f. 22}

I now took leave of Mr. Grant to whose extensive local information and patient willingness to resolve any difficulty, amidst the pressing duties of factor on an extensive estate and principal magistrate in a populous district, I consider myself very much indebted. Proceeded forward to Aviemore.

I sent for a guide to show me the southern boundary of the Parish of Duthel, viz. Lewis Grant of Lethendy A… [undecipherable], ground bailiff on this part of Sir Ja^s^ Grant's Estate. Meantime I forded the river to Rothymurchus to call on Capt Cameron at ----[blank], factor on that estate who pointed out to me the line of march with Aberdeen. This line being very simple little more was necessary than to have several names and a description of it. From the east side of Cairn Goram the boundary still runs according to the fall of the water SW to Brea Riach, thought even higher than Cairn Goram.[22] Between these two mountains is a deep and narrow but tolerably level glen, from where flows one of the principal springs of the Dee; and towards the Spey we have the Alt Dru, which being joined by the Benny from Loch Errich and the water of Loch Morlich, forms the Druie, on which a number of saw mills have been constructed for sawing the wood of the extensive forests of Glenmor and Rothymurchus. On the east of Cairngoram we have the source of Nethy within a few hundred yards of Loch Awin.

From the Carn Goram the march between the D. of Gordon's forest of Ben Awn in Banft and the Earl Fife's property in the head of Dee in Aberdeen runs east again according to the fall of the water, by Bein Mach Dhubh etc.

{f. 23}

From Brea Riach by a comparatively low ridge, the boundary of Inverness passes on towards the south to the top of Scarsoch where it meets Perthshire.

To perambulate this part of the boundary would be impracticable. It was fortunately unnecessary. I had indeed intended to visit the mountain Scarsoch being the junction of these counties by following the course of Feshie. I soon found that all the information I could thereby obtain would by no means recompense the time and labour. Before leaving this however I may mention that should any road be thought necessary from the vale of Strathspey eastward, e.g. to the Spital of Glenshee, the vale of Alt Dru seems likely to afford the easiest summit, situated between the mountains Brea Riach and Cairn Goram. It seems extremely deep and narrow, but in every other quarter the mountains present us with the most stupendous precipices.

There is a good turnpike road from Aberdeen for nearly 10 miles up the north side of Dee. A continuation of this would open to the east the country of Strathspey and Lochaber.

There is at present a track from Blair in Athol by the river Bruar, and the mountain ranges northwards. Travellers generally descend into Strathspey by the River Feshie. The post to Ruthven Barracks used formerly to come this way from Perth, descending by the Frommie. This track is marked on the map as far as the bounds of Perthshire, and I have heard that it is intended to complete the military road that way, by which nearly 20 miles would be saved. But Mr. Arrowsmith does not seem to know that the great highland road is that along the river Garry to Blair and so by the rivers to Perth.

{ f. 23 verso }

Craig elachie, qu? Craig dhelachie (dh is scarcely sounded or approaches to the sound of *y*)—the rock of division or separation. This is the motto of the clan Grant & was formerly the cri de guerre or war hoop of the tribe.[23] There is another Craigelachie of less magnitude & note on the Spey 40 miles lower, where the parish of Knockando bounds with Rothes. The intermediate district is the proper Strathspey.

The sea visible here is evidently that at Findhorn 33 miles distant but as it is not probable that we could see the immediate shore on account of the intervening ground let us suppose 36 miles only. This will not give us an altitude of more than 1000 feet which is surely far too low for this mountain as there was a patch of snow just beside us. Could we suppose this the point of congelation our elevation according to Mr. Kirwan[24] would be nearly 5000 feet which is certainly far too high. We will probably come much nearer the truth by considering that we looked over the high ground near Dunduff whence the shores of Sutherland are pretty distinctly seen. This would give for our utmost horizon a distance of 30 miles and an altitude of 2200 feet which I dare say is pretty near the truth. I am thus particular with respect to this mountain as it may afford a means of estimating the height of the mountains of Monaghlea.

{f. 24}

Wednesday June 18th Aviemore.

Upon returning hither I found Lewis Grant waiting for me with whom I immediately proceeded to perambulate the neighbouring part of the boundary. Being the 'march' between the property of the Duke of Gordon in Badenoch and of Sir Ja^{s} Grant in Strathspey, this boundary is sufficiently precise and is even marked by stones. We had a shepherd from the neighbouring farm in Alvie parish to accompany us. Accordingly setting out from the side of Spey, we ascended the steep rock of Craigellachy [*sic*] which forms the principal feature of this boundary. After this we had a long walk by the burn of Lyne Builg to the summit of the Mountain Ceal Churn Beg from where there is a very extensive prospect over Strathspey and the vale of the Dulnain. Bel Rinnis appears N85E p compass, Bynag S55E, Cairngoram and Brea Riach from SE to S, a long distant hill in the N20W , Lochindorb N50E and the sea beyond it visible.

The boundary descends directly to, and crosses the River Dulnain, ascending from thence to the summit of the opposite mountains and running NE by these summits to the outline of Nairn near Loch Bruich.

Having taken a sketch and description of this boundary I descended to Aviemore and prepared to prosecute my journey southward. I had now finished the whole eastern boundary of Inverness pretty much to my satisfaction. From its peculiar intricacy I had found it necessary to enlarge the scale of that part of the map in order to be enabled to delineate the boundary with tolerable distinctness.

{f. 24 verso}

Monaghlea or that desert tract of mountain extending between Badenoch & Stratherric. The hill upon which I stood formed one of the skirts of it; part of it lay immediately before me but to the west several neighbouring hills obviously higher than that whereon I stood quickly cut off the prospect. I cannot therefore suppose the average height of the Monaghlea west of Dalnalealg to be less than 2500 feet above the level of the sea & since there are few or no gaps or passes in it, or at least none that intersect it to a considerable depth, I do not think any good or convenient communication can be established across it. In all roads to the north therefore we must look either to the valley of the Spean on the west of the Monaghlea or to the neighbourhood of the Spey etc. on the east of it. The road to Fort Augustus indeed goes directly across this, but everyone knows this to proceed literally by the summit of a mountain.

{f. 25}

The peculiar form of the county of Moray may, I think be partly explained by attending to the following circumstances.

1. There was a sheriff, or 'vice comes' of Elgyn at a very early period, but his jurisdiction does not appear to have extended over the whole of the present county.
2. The greater part of the lands in the south of Inverness and Moray, belonged in the 15th Century to the family of Comyn, who were afterwards proscribed and forfeited during the reign of Robert the Bruce on account of their opposition to that Monarch's claim. Robert erected the whole country between Spey and the western ocean into a comitatus or Earldom, bestowing it on his eminent partizan Thomas de Randulph who was thus first earl of Moray. The charter is still in existence.
3. Upon the deaths of the two sons of this Thomas, without issue male, the Earldom anno 1346 reverted to the Crown. But Patrick Dunbar, Earl of March, having married Agnes, daughter of Earl Randolph

[*sic*], was usually stiled [*sic*], Comes Marchiae et Moraviae. Their son John Dumbar [*sic*] having married a daughter of Robert II, was created Earl of Moray, with the exception of the lands of Badenoch, Lochaber and Urquhart and according to Boece [*sic*], of Badenoch, Lochaber, Petty and Brathla qu. Brackla?, and I think 'tis probable that more may have been excepted.

In this family the Earldom continued for two generations. James Dunbar [*sic*] the last Earl married his relation and had a son Alexander, but the mother dying before a dispensation was obtained, the power of the Douglasses got the Earldom transferred to a daughter by a second marriage, who had espoused one of their own relations. The Douglasses afterwards forfeited the Earldom by rebellion in ———[blank].

Alexander, who had been thus deprived of the Earldom and whose abilities are much celebrated by Boethius lib. XVIII[25] was however knighted and made Heritable Sheriff of Moray. This office remained in his family till after the Union. The barony of Abernethy made part of the earldom of Moray and gives still the title of Lord Abernethy to that Earl. The Grants possessed the lower part in the 16th Century and afterwards purchased the upper part from the Earl Moray for which a feu duty is still paid.

{f. 26}

I have already stated that Badenoch was excepted from the second Grant of the Earldom of Moray. King Robert Stuart granted this to his son Lord Alexander commonly called the wolf of Badenoch, who being excommunicated for keeping forcible possession of the Bishop's lands, burnt the Cathedral of Elgin. After his death the Lordship was possessed by his brother Lord David, upon whose death it reverted to the Crown.

The Earl Huntly having quelled a dangerous rebellion at the battle of Brechin 1452, was rewarded with the Lordships of Badenoch and Lochaber, which have remained in that family ever since.

The farms or manors of Gartenmor, Bymore and Tulloch in Abernethie [*sic*] and Tullochgoram, Clourie and Cour in Inverallan, were originally considered as part of the 'sixty davochs of Badenoch', and hence we may account for them being a part of Inverness Shire. They were exchanged by the family of Gordon, for lands belonging to the family of Grant in Banffshire. Shaws Moray.[26]

I have not been able to discover how Cromdale forms part of Inverness. It is described as situated in that county in an annunciation by Isabel Countess of Fife, to Robert III. Sibald's [*sic*] Fife.[27]

There were afterwards Thanes of Cromdale, from whom the Grants of Grant purchased this property.

{f. 27}

Friday June 20th

I set out from Aviemore for Pitmain. It may be worthy of remark that there is a marked improvement on this part of the military road. Leaving the old line at Kincraig, it proceeds down to the side of Loch Inch along the edge of which, and near the river Spey, it runs on without the smallest pull to Pitmain.

As I found myself now at a loss for information and as I had already reaped so much benefit from the plans of various estates, I called on Mr. Macpherson of Belville, who had lands, I knew, in the valley of the Truim, and possibly near the county boundary. Mr. Macpherson (son of the translator of Ossian) who seems much interested in the welfare of the Highlands, immediately showed me the plans of his lands, from which however, as they formed no part of the boundary, little useful information could be derived.

It may, however, be observed that the lands of Belville and all others on the north bank of Spey here, run back across the head of Dulnain, until we meet the feeders of the Findhorn River in the Monaghlea.

The upper part of the valley of Dulnain is desert, the only habitation in the Inverness part being Caittin, in the immediate vicinity of the boundary. The remainder forms part of the pasture lands of the farms on Spey.

Mr. MacPherson proposed that we should immediately proceed to a meeting of the Justices about to be held in the village of Kinguisie, where it was probable I might be directed in the further prosecution of my journey. I had intended to cross Spey and visit the minister of Kinguisie and Inch but was informed I would see him at the meeting.

{f. 28}

Kinguisie is a neat little village situated on the high road from the parish church towards the Inn of Pitmain. It has chiefly arisen within these three years under the auspices of the Duke of Gordon and is likely to become a thriving and populous town.[28] It has already a post office and attempts are making to introduce some of the coarser branches of the woollen manufacture. The houses are built with stone according to a plan and some of its buildings may be even styled elegant. It possesses several good falls of water which are well employed already but are eventually likely to become of much greater importance.

Kinguisie is the only village (if we except Grantown from which it is a distance about 25 miles) between Dunkeld and the shores of the Moray Firth. From its situation it seems well adapted for an inland manufacturing town and by its rich meadows and arable fields on the Spey, affords a pleasing anticipation of what the Highlands of Scotland may become, when once laid open with good roads and the industry of the people directed to useful and valuable purposes.

Kinguisie derives great advantage from the great military road to the north running directly through it and one of the stages on this road is hard by the town. Much benefit is also expected from the road which it is proposed to carry on by Loch Laggan to Fort William.

{f. 29}

From the meeting of Justices I could derive very little information. The only person present who was at all acquainted with the boundary on the south of Inverness Shire was a Mr. McPherson of Ralia, a gentleman near 80 years of age. Mr. Anderson, minister of Kinguisie, proposed that I should call on certain shepherds in that quarter who would point out the contested boundary in the hill of Drumochter, but as this was the season at which they collect their sheep for shearing, it did not appear likely that I would readily find these men. The only source of information seemed to be Ralia whom I was at length so fortunate as to induce to accompany me to the spot, and with this view we proceeded that evening to his residence of Breakachy on the road from Dalwhinnie to Garviemore.

Proceeding from Pitmain up the Spey we have on our right the tremendous precipices of Craig Dhubh (pron. Craigoo) or the black rock, perhaps the most remarkable on the banks of the Spey. The height may be near 1500 feet above its base, almost a continued precipice rising immediately from the road. As we ascended the vale of Spey opens into a wide and level valley, the cultivation of which is however prevented by the frequent and extensive inundations.

Saturday June 21st

Proceeded with Ralia by Dalwhinnie to the boundary in the hill of Drumochter. Ralia has been for many years the Duke of Gordon's tenant on the very lands in question. He is himself the principal evidence on that side and having

{f. 29 verso}

Dalwhinnie Inn. Saturday June 20th. Attempted an observation and although it appears of small value I set it down viz.

XII 5' 5" D. Alt 112
40' 35"
Too late for merid
XII 12 50
24' 20"

14 25
18' 10"
14 25
15' 40"
16 35
11' 20"

S. Diam fore 0 51' 40"
Back 9 48' 20"

The day very cloudy. Error by Mth [?]
20' 20"

Torc—the boar—from some fancied resemblance.

{f. 30}

been a party at many attempts to settle the dispute, is equally well acquainted with the claim of the Duke of Athol. I consider myself therefore as personally fortunate in having the statement of this matter from himself upon the very spot.

The marches to the eastward follow exactly the summit level and are therefore sufficiently distinct until we come to the summit of the mountain Buinac. This summit is broad and flat and is claimed by each party. The Duke of Athol's tenants propose to follow still the summit level, or fall of the water, across the military road to the middle of a hill named Torc and so to the summit of Ben Udamin,[29] near Loch Eruch,[30] claiming in fine the whole tract drained by the feeders of the Garry.

Immediately south of Torc, we have a little valley named Corry Dhon (pron. Ghone) the waters of which running directly E. and meeting a stream from the opposite hill called Ault a Churn, form the head of this branch of the Garry. These two corries are claimed by the Duke of Gordon's tenants on the footing of prescriptive possession, and with some part of the neighbouring mountains to form a regular outline, form the principal basis of the dispute.

The breadth of this tract measures at the military road, 1100 yards. The length following the marches from Ben Udamin to Buinac may be 6 or 7 miles. It is scarcely necessary to remark that being on the ridge of the Grampians, the very 'Domum Britanniae' there is no habitation nearer than 5 miles.

{f. 31}

Saturday—Loch Eroch side

After dining at Dalwhinnie with Mr. McPherson Ralia, we proceeded along the side of Loch Eruch. We were flattered for some time with the appearance of a path, but it gradually disappeared and the mountains hanging precipitously over the lake we had some difficulty in getting forward to a shooting quarter belonging to the Lord Chief Baron on the west bank named Dal n' Lynchart,[31] where we remained that evening.

The shooting season not having begun, there was nobody in this solitary dwelling, but one man who with his wife looked after the house and garden. I found here Stobie's map of Perthshire.[32] He takes no notice of the Duke of Gordon's claims. As my map begins to run out, I took a sketch of the outline so far as the county of Argyle. Stobie however, is not to be compared in point of accuracy or minuteness with the skeleton map I have, although this on a far smaller scale.

I had been directed to Mr. Mitchel's hinds (shepherds) on the water of Ailler for this part of the boundary, and found it therefore necessary still to push on by the lakeside, although we found considerable difficulty in getting our horses forward among the rocks and stones. There was indeed no choice of road between the mountain on the one hand and the lake on the other.

Loch Eruch is very deep. It seems the gentlemen at Dal n' Lynchart have found soundings with 80 fathoms.[33] It never freezes excepting for a little way at the upper end where it is very shallow.

{f. 32}

Ice is equally unknown in the River Eruch until it falls into Loch Rannoch. The great depth of the lake, which the appearance of its banks abundantly confirms, is evidently sufficient to account for this circumstance which I find is by no means peculiar to Loch Ness and Loch Lochy alone. Loch Tay, Loch Maree, Loch Morrer etc have never been known to freeze, even in severe winters, as in fact we have never a frost of sufficient duration or intensity to cool all the strata of such bodies of water.

It is affirmed of the lake Duntelchag in Stratherric that altho it is not known to freeze in the early part of winter, that one severe frosty night in the beginning of Spring will be sufficient for that purpose. It is not difficult to give an explanation. In the winter while the cooled strata subside from the surface they are again warmed by the contiguous earth. The process of cooling will go on therefore very slowly and in our changeable climate may meet with many interruptions insomuch that the waters of the lake may not reach the temperature 40 degrees or lower for two months. But if then a frost should take place, this lake will freeze as readily as any other water.

Some experiments I have made on Loch Ness[34] induce me to think that the great body of its waters never vary much from the medium temperature of the Earth, but they are considerably cooled in descending to the sea—a property similar to that of Loch Duntelchag and remarked of Loch Lydon in Rannoch and no doubt of many others in the Kingdom.

{f. 32 verso}

L. Eroch side. Sunday. Opposite Bein Udamin.

Mer. Alt. 56 47' 25" doubtful. From waters below Aonach Mor distant 1 mile and sitting at edge of lake gives a dip. probably of 3 to 4 min Lower limb

XII 14 22 D. Alt 112 22 30
XII 34 10 111 13 -

S.D. fore 9 47 50
R 10 50 -

The station was a mile S of Camus na Ghrain & may be taken as the lat pos [?] S limit of Inv[s] , E of Loch Eroch.

{f. 33}

Pass a shepherd's hut called Camus na Ghrain. The men all in the hills about a mile to the southward. We are opposite the Burn of Ben Udamin or Ben Udlamin, the march on the east side of the lake. I endeavoured here to take an observation for the latitude as stated on the other side.

It was not without some danger that we got our horses forward among the large stones and rocks, around the base of Ben Aillar or Aular. At length however, we arrive on the water of Aillar, the southern boundary. The shepherds here, to whom Macpherson of Clurie had referred, but without giving them previous notice, were all busy gathering their sheep in the hills. The only information therefore to be had was that the neighbouring stream Uisgeaillar was the march of this farm (and consequently of the County) from the lake up to the first fords, that from thence it passed between them by march 'cairns' to the westward. This information turned out afterwards to be perfectly correct.

This like many shepherds huts is placed just upon the bounding stream of the farm. The principal occupation of the possessors is to prevent the cattle and flocks from trespassing. Their knowledge of the common boundary is obviously unquestionable. I had it now in my choice to return along the lake or to endeavour to proceed to Rannoch. Our journey by the lake had been so troublesome that I chose the latter. The country was now quite open and comparatively low. The distance was stated at only 5 miles to Loch Rannoch.

{f. 34}

Sunday June 22nd P.M.

Cross the Aillar or Aolar and enter Perth, we soon find ourselves deceived with the route we had chosen. The ground is one entire moss, not the least track visible. I certainly had felt that Mr. Arrowsmith ought to erase the line he has drawn by the side of this lake, yet being in want of any guide (for I had mislaid the sketch from Stobie) I would have given anything for this corner of the map to direct me. We arrived at the end of the lake in about 4 hours though the distance can hardly exceed 2 miles. We found much difficulty in crossing a little black stream which falls into the lower end of the lake, on account of quicksands. At length however, by recollecting that the only way of crossing streams of this kind is by fording into the lake, [we] got safely over.

The view of the lake, Ben Aalar etc at this place has much of the sublime. The mountain is upwards of 900 yards above the level of the lake. Aonach mor opposite is at least 600, Ben Udamin probably 700.[35] I took these altitudes at Camus na Ghrain, guessing the distances by the help of the map. This place would have afforded a better opportunity of procuring them but we were too interested in getting forward to some habitation before nightfall to stop for such a purpose.

There is a singular tradition that Loch Eroch was formerly a parish, which has now submerged in the lake. Part of the crofts are said to be visible running down to the water. I should rather imagine that they are merely the 'detritus' of some ancient mountain streams. The lake however, is a singular object. It forms a complete gap in the general chain of the Grampians. The upper end is almost level with the Truim and I do believe that were the Landlord of Dalwhinnie to cut his pents with that view, he could in a short time send all its waters to the Spey.

{f. 34 verso}

Rannoch. The Ferny. The country abounds with the fern plant.

{f. 35}

We arrive on the bank of Loch Rannoch at ten at night and fortunately the nearest house was possessed by Mr. Menzies, factor for Mr. Rob[t] Menzies on the estate of Rannoch. Upon introducing myself was very kindly received by Mr. Menzies and his son Mr. ——[blank] writer in Perth. They were here at the sheep shearing. They explained the cause of the proceedings in Perth. The committee appointed to attend to the business of surveying the boundary were His Grace the Duke of Athol (who was at the time in London on the impeachment trial), the Earl of Breadalbane (also in Parliament), and Sir Rob[t] Menzies (who was at Bath). This was indeed a singular contrast to the guide proposed by Macpherson of Cluny viz. an old Highland shepherd on the water of Aillar.

Monday, June 23rd

Our horses having lost two shoes on Loch Eroch side, I sent them down to the head of Loch Tumel to be replaced. Meantime Mr. Menzies Jun[r] and a guide named ——[blank] Menzies proceeded with me to a neighbouring hill and pointed out the march from Ben Aillar to near the head of the Blackwater or Levin. The greatest part of the same line I afterwards examined more minutely with a guide. The marches have been settled and 'march cairns' or stones placed which prevent any dubiety in this quarter.

I called this evening on Col. Robertson of Strowan on the S of the Gaur Water who had also been appointed of [*sic*] the committee on the subject of boundaries. They appear to have expected here that I would delay proceeding in this business until the arrival of the Duke of Athol. Indeed they seem to have considered the object as of much

{f. 36}

greater consequence than it really is and as being in fact a definitive settlement of the county boundary. Of course no one chose to interfere with the property disputed by the Duke of Athol and there the matter rested.

I have had some conversation on the subject of roads. They do not expect, in this quarter, much benefit from the proposed road from Lochaber to Killin, which of course must run through this quarter. It will be chiefly a drove road and consequently beneficial only to the northern counties. A great part of this road will pass through Perthshire and being, after all, only along the extremity of the county in a thinly inhabited district, it can hardly be expected that the county of Perthshire contribute willingly to its execution. On the other hand a great part of the military roads to Fort Augustus and Fort William lie in more populous districts the funds of which might perhaps without impropriety be diverted to an object of this kind.

What would be of more consequence to the District of Rannoch is a road from the head of Loch Rannoch to the western ocean at Balhulis [*sic*]. There is not at present a road across the central Highlands from the line of the Canal on the north[36] to the Strathlay and Blackmount on the south. The whole intermediate country can only communicate with the sea at Perth or Inverness by a distant and expensive land carriage. The proposed line through Glen Spean by Loch

{f. 37}

Laggan side will be of essential service in this respect to the northern part of this district. The southern parts and Rannoch in particular will be in the same situation as before. They are now at the distance of 70 miles by county roads from Perth and yet are only 27 miles from the sea. Nay, of this there is only 18 miles of road necessary along a level valley with hardly any streams, in order to communicate with the military road at Kingshouse from whence a road runs down Glenco to the arm of the sea called Loch

Levin. Towards the east there are good county roads from the head of Loch Rannoch to the military road at Tumel bridge from whence there is a direct communication to Blair, to Dunkeld, to Perth, to Crieff and Stirling etc. In this way the whole of the counties of Perth, Argyle etc would be laid open to Fort William and the west Highlands by the shortest and easiest course and a complete communication formed across the center of the Kingdom from the Eastern to the Western sea.

I was just about to cross this very district and had no choice but to send my horses to Kingshouse by Glen Lyon and Loch Tay, a round of 90 miles, or to carry them as I might across the moor. I adopted the latter, but indeed had not the weather been uncommonly dry for some time before, I must have certainly lost them. I was several times in danger of doing so.

{f. 38}

Tuesday June 24th

Returned across the water of Gaur to Mr. Menzies's. A guide had been procured to accompany us by the muir and side of Loch Lydon to Kingshouse. This man had been one of the principal evidences in settling the marches of Rannoch and was of course well qualified to give any necessary information on that subject. The morning proved so wet and cold that we had some doubts whether the muir would prove passable. We set out however about 12, still rainy. Pass the farm of Dunan, a few feet of rise brings us to the summit level of the country. We have then an extensive flat valley, or rather plain, almost covered with lakes. The first of these, Loch Aich (misty lake?) is not extensive but has on its banks some beautiful meadows carrying excellent crops of natural hay which is piled into stacks on its shores by the sheep farmers of Dunan etc.

An ineffectual attempt was made some time ago to drain this lake by a cut through a rock near its mouth, with a view to increase the hay meadow and preserve it from floods. The scheme is certainly practicable but from the hardness of the strata it probably proved too expensive.

I can have no doubt however, from what we have seen on the banks of this lake, that the shores of Loch Lydon etc. etc, nay maybe perhaps the greater part of this moor, at present entirely covered with an unprofitable moss and exhibiting only a wide and desert waste, might be converted into hay pastures as luxuriant and perhaps as valuable as any in the Kingdom. Then only can we expect that the neighbouring extensive tracts of hill country can be completely depastured, for however extensive

{f. 38 verso}

Cruach. The Stack. When viewed endways it has a considerable resemblance to a hay mow.

There are several hills of this name, one in particular of great magnitude on the bank of Loch Awe.[37]

{f. 39}

the districts may be that are fit only for pasturage, the flocks must necessarily be limited by the quantity of 'wintering' in the power of the farmer.

Hence we may account for that species of policy which induces the sheep farmer to remove as many small tenants as he can. The spots which they inhabited are green, having been long under the plough. They are not liable to 'storm', being usually situated in the bottom of the valley. They form of course his most valuable winter pasture. Hay having been already cleared, he may lay much of them under the plough and raise artificial green crops besides. He removes neighbours who may be troublesome to him but whom he does not expect ever to become useful. Yet I cannot think the sheep farming system must necessarily be unfavourable to population. Nay in the very instance before us it is certainly the reverse. On the north side of the moor are several houses inhabited by shepherds where there formerly were only wretched shealings for the shelter of herds during a few weeks of summer.

The south side of the lakes however is a mere desert, being an entire bog, perfectly impassable. It is quite destitute of trees. The north side on the other hand is pretty well wooded, but they are going quite into decay. The trees in the western parts of the Cruach (a hill district on the north bank of Loch Lydon) (pron. Leytan) exhibit a ruinous picture. Numerous trunks lie half rotted on the ground, serving as dams to obstruct the hill torrent and lay the foundation of a future moss. There is little or no valuable wood standing. The transportation is scarce practicable.

{f. 40}

With one of the neighbouring shepherds, or rather sheep farmers, for the person named Jeffrey was one of a small company who had taken a lease of the south side of Cruach as a sheep walk for 400 Guineas. The present rent is scarcely a fifth of this sum.

With this man and my guide Alex Cameron from the mouth of the river Eroch, I examined the junction of the counties of Inverness, Perth and Argyle, at a little pool named Loch a Chlai, forming the source of the Uisg Dubh or Blackwater which as it increases after traversing several lakes, receives the name of Levin. The south side of this water is extremely boggy. The pasturage improves as it ascends to the hill of Cruach which divides this little slip of Argyle from a similar slip of Perth, interjected between this hill and Loch Lydon.

Loch Lydon is at least 10 miles in length, as beautiful a sheet of water as I have seen. The north shore is much wooded although now in decay. The south, as already mentioned, a flat bare impassable morass. The west end of the lake is full of islets and much indented with bays, the most considerable of these (and which gives to the whole somewhat the appearance of the letter Y) runs to the N.W. and forms the boundary with Argyle and Perth.

A continuation of this is named Loch Badan Choig (Hawks bush) and receives into its head two small streamlets the most eastern of which forms the limit of Perth Shire.

To the S and West of Loch Lydon a cluster of small lakes seem to cover the greater part of the moor. On the south the mountains of Glenurchy rise like a bounding wall. The same appearance is presented to the West by the steep hill of Carn Dag, the surface of which is streaked with snow.

{f. 40}

'Luib' The bight or loop formed by the winding of a stream or at the junction of two streams. More particularly the flat green pasture usually found in such situations this having been a resting place on the drove way immediately before ascending the black mount.

{f. 41}

When we came near the western limits of Perth our horses unexpectedly sunk in soft ground to the belly. In struggling to get free, the pad on which my instruments were fixed was broke, without however any other bad consequence to them. The servant who rode that horse got a slight bruise. The men carried the instruments and we led the horses on to Kingshouse. As we enter Argyle we find a sort of path which winds in a strange manner, but must be adhered to. Ground solid but extremely stony and uneven, the whole tract to the southwest as far as the black mount covered with bogs. Many roots of trees in every direction, but not one standing trunk to be seen.

We descend by the bed of Ault Etie to Kingshouse where we arrived late in the evening. Kingshouse is an inn built by Government on the military road to Fort William. Many of these inns are so named (Tui 'n Righ) by the Highlanders. The spot has the more appropriate name of Laop na Marst (the bight of cattle). The house is in very sorry condition and cannot be said to do much honour to his Majesty. It is indeed a solitary dwelling and unfortunately the possessor has no 'farm'. Some patches of potato ground however, scooped out of the moss, show us that much might be done in the agricultural way even in this bleak district of the Highlands. Arthur Young I think has said that no improvements are as valuable as those made on a bog, but the farmers and proprietors of the Highlands have this, in a great measure, yet to learn.

{f. 42}

The large valley we had traversed terminates to the west in a singularly abrupt rock named Buachal Ety (the herdsman of Etive). Overhanging the Kingshouse, its perpendicular crags give it the appearance of a cluster of basaltic columns. It divides Glenco from Glen Etie? [*sic*]. These two openings appear from the east like two narrow alleys among the houses of a street, to compare great things with small. The district between them forms the Royal Forest of Dalness where two or three hundred deer may sometimes be seen grazing. Many names of physical objects in the neighbourhood bear allusion to the deer, or at least to the subject of hunting, a thing common in many parts of the Highlands, thus Ault na fedh (pron. Fay)—burn of the deer, Meal 'n Ruac—deer hill, Loch Eilt—deer lake etc.

Wednesday 25th

Left Kingshouse and proceeded down Glen Co. The old road to Fort William rises up the hill to the right by what is called the 'Devils Staircase' and by the Highlanders Mam a Grianan (sunny path). It passes the River Levin at Kinloch and afterwards crosses two other mountains (Mam mor and Mam beg—the great and little path) before reaching Fort William. On account of these disagreeable ascents the road is now nearly deserted (and in consequent bad order) for the road down Glenco and by Ballahulis. The hills on the south of Glenco may be termed the Montserrat of Scotland, being full of naked peaks and perpendicular craggy chasms. Down these the various streams, at all seasons but particularly during thaws, pour immense torrents of

{f. 43}

loose gravel and stones, covering the whole valley with this destructive kind of flood. The road in particular has been in many places strangely disfigured by it. 'Tis by no means uncommon for a bridge over one of these streams to by literally buried several feet under the surface. These torrents of stone form perhaps the most striking feature in Glenco. They are indeed well known over every part of the Highlands, but I have seldom seen them so numerous or so vast as in this quarter. It must appear a singular scene to the traveller to see whole mountains as it were in motion, or going to wreck. I should imagine it little inferior to the avalanches of the Alps and as wanting only fire (the most terrific part without doubt) to have all the destructive appearance of a torrent of lava.

The falling of these stones is often dangerous to people passing beneath. Several have been wounded in Glenco. It is not confined to seasons of rain, but occurs at all times in a greater or less degree. The day on which I passed through Glenco was very fine and warm, yet I heard the trickling of stones, so to speak, in various places of the mountain beside me.

An interesting story is told of a cottager on the bank of Loch Awe who, some years ago, when sitting in his hut which was situated at the bottom of a steep mountain, heard a near and crashing noise, which upon peeping from a window he saw was occasioned by several detached masses of rock bounding down the hill. One child stood at his knee, another was sleeping near him. His wife had just gone out and he heard her cries. He sprung to the door

{f. 44}

with a child as he thought in each hand. An enormous stone passed at that instant through his little dwelling and laid it in ruins. He missed one of his children and no longer heard the cries of his wife. Upon looking however to the remains of the hut, he found his child in a corner whither it had been thrust by the fragments displaced by the rock in its passage, alive and unhurt. The mother immediately coming to the scene, their joy and gratitude may well be conceived.

A similar accident but having a more melancholy catastrophe occurred lately at Griban in Mull. There had been a wedding in a small cottage situated under a mountain of the above description, when after the usual enjoyment, the new married couple had been put to bed and the wedding guests had just departed. A tremendous rock descended from the hill upon the dwelling of the young pair and buried them in its ruins! The stone is pointed out by the neighbouring cottagers under the name of Clach na ———- [blank], the stone of the nuptial bed.

It seems these land slips have increased much since the hills were de-pastured with sheep. On purpose to improve the pasture, the heather and young wood whose roots bound the loose soil of the hills have been all burnt off and eradicated while the sheep form tracks and roads on the sides of the mountain, which serving to convey the water of a considerable tract to one point, form falls that cut upon the loose soil till at length the bank slips down and sets many tons of this loose friable stuff into motion. There is an appropriate Gaelic phrase for this. The Highlanders called such a slip a 'Scridan'.[38] It will always form a great obstacle to the

{f. 45}

complete preservation of roads conducted through a country of this description.

The lower part of Glenco is beautiful and romantic, it is also pretty well cultivated.

I called here on Mr. McDonald of Achriachtan[?]

with whose family I was a little acquainted. I found there Langlands' Map of Argyle,[39] but it does not appear to be at all accurate in the borders of Rannoch and Lochaber.

Mr. McDonald pointed out to me in Loch Levin a small island or rather three distinct islets called Elan Mhuin, or St Mungo's Island. It is not now inhabited but there is on it the ruins of a chapel. It is used as a burying ground common to both sides of the Loch. The pasturage brings about £25 per ann, which is alternately paid to the proprietor of Glenco, and to the proprietor of Callart on the opposite shore. This Isle therefore must be considered as belonging alternately to the Counties of Inverness and Argyle. Two other little isles in this inlet belong to the estate of Callart and of course to Inverness.

Pass the slate quarries of Ballahulish where I had my horses shod. They had lost some shoes in Rannoch Moor and were almost lamed in descending Glenco. These quarries afford employment to nearly 300 people, whose cottages and little patches of cultivated ground interspersed among the woody banks of this inlet give a very enlivend [*sic*] appearance to the scene. A good deal of craft frequent this Loch for the slate and some lime which is wrought to good account by the proprietor. The interior of Loch Levin is

{f. 45 verso}

Ballahulish Thursday June 26th

XII 13 -		112 48'		
10"	Sol. D.	9 47 20		
XII	16 30	112 41'		
10"		0 52 40		
1st mean	XII	14 45	112 44' 40"	40
Index error 20				
XII	27 20	112 - -		
XII	36 -	111 48' 50"		
2d mean	XII	31 40	111 54' 25"	
3d	XII	42 45	110 42' 20"	

The time requires correction as usual.
Day bright and hot. A north wind.

{f. 46}

very deep. There is a shoal at the entrance on which however there is always 5 or 6 fathoms water.

Arrived in the afternoon at Mr. Stewarts of Ballahulis, situated about half a mile south of the ferry, where I found Mr. Chas Stewart who had formerly agreed to accompany me.

The prospect in every direction showed nothing but steep peaked mountains, the sloping sides of which descend immediately down to the sea. Their height however, is in general not great. Few of them seem to exceed 1500 feet.

Thursday

The annexed observation may afford the latitude of Ballahulish. It agrees well with the map.

I had an opportunity here of comparing Langlands' with General Roy's map.[40] The difference in the position of the Locheil and Linne dheloch is very material in so much that I find a sketch from Langlands of the northern part of Argyle cannot be adapted to my skeleton map. The Nevis will confound itself with the Loch Levin. I fear Langlands took the body of his map from Ainslie and that his survey consisted only in filling in the minutiae in the best inhabited parts of the county from an eye draft. As a proof of this the boundary with Inverness is a river and chain of lakes connecting with a lake called Loch a Chlai (Sword Lake) or in his map Loch 'n Altich. No notice is taken of this water in his map and Loch Lydon, although nearly true in shape, has little more than half the real length.

{f. 47}

Friday June 27th

I rode down the coast of Argyle in company with Mr. Stewart Junr of Ballahulish. We proceeded only to the Ferry of Connel and returned. The view down this arm of the sea is certainly extremely pleasing, the surface of the water covered with beautiful green Isles and the fertile island of Lismore occupying the middle of this bay was well contrasted with the lofty peaks of Mull immediately beyond it. A great part of this coast is wooded. The wood is chiefly kept in coppice for the sake of the bark and charcoal, under which mode much of it will afford a regular rent of 20 shillings an acre. Pasturage however and the culture of potatoes seem the chief object of attention.

The roads along this shore are in excellent order. Much indeed has been done by the County of Argyle in this respect. Yet I am told the isles, Morvern etc, are still much in want of this convenience.

The most troublesome circumstance in travelling along this coast is the frequent ferries. This however is unavoidable and from being intersected by numerous arms of the sea, Argyle surely enjoys many advantages. Like natural canals they have brought every part of this interesting county within easy access to water carriage, to fishing and the many other advantages therewith connected. These however, are even yet little improved and excepting at Campbelltown, the county of Argyle had until lately, with 1500 miles of coast, not a single vessel of burthen.

{f. 48}

Saturday June 28th

Left Ballahulish and arrived at Fort William. At the house of Glen Nevis I found Mr. Angus Ranken, tenant of Kinlochbeg on the Levin River and forester of the Royal Forest of Dalness, by whose assistance I was enabled to complete my sketch of the Levin River, Cruach etc.

The River Levin, as will be seen from this sketch, passes through a chain of lakes of which the uppermost named Lochan na Bhaillie (pron. Vallyie) may be nearly three miles in length. Two islets in it viz. one near the upper end and another near the middle belong to Argyle. Below this a chain of 4 or 5 smaller lakes forming in a great measure one general lake for 2 miles receives the name of Dubh Lochans (pron. Doo) or black lakes. Nearly two miles below this and of equal length we have the third Loch 'n Inner (pron. Eenyer). The river now falling over several cataracts, the lowest of which named Is n' Smoudie (Smoaky Falls) is of considerable magnitude and receiving from the north the branches Ault Eilte from Loch Eilt or Deer Lake and another from Corry na Ba (or the vale of the Cow) falls into the head of Loch Levin.

The country to the north of the Levin has the name of Mamore, forms part of Lochaber and by an old Act of Parliament is included in the Shire of Inverness. The outline of this county proceeded therefore by the sea shore to the mouth of the River Lochy, which it ascends a little way. The northern shore of Locheil, by some singular arrangement is included in the county of Argyle. An island of some magnitude in the Lochy, between the Castle of Inverlochy and the sea belongs to Inverness.

{f. 49}

Monday June 30th

I had learnt that Mr. Cameron of Fassifern was the best informed person in this quarter respecting the boundaries of Argyle and Inverness. I therefore prepared to cross the river and wait upon him.

On my way I called on Mr. Stewart barrack master of Fort William, who with Capt. John Cameron of Kinloch Arkaig confirmed the accuracy of the sketch I had made of the River Levin etc, that district being well known to both of these gentlemen.

The day proving excessively rainy I stopt some time at Corpach with Messrs John Telford and Donkim.[41] I had in the mean time sent notice to Mr. Cameron and though the rain had not abated, set out towards evening for Fassifern. A considerable part of the Loch na Gaul road having been formed and the bridges were many of them built altho not covered with earth. I felt however, no great inconvenience from this although the streams were high until I came to the water of Fassifern. This had the appearance of a mighty river. The night was dark and no one was near me. I rode a spirited little Highland pony and made two ineffectual attempts to ford the stream. At length however, by discovering a place where it spread to a great width on the old road near the shore, I got safely over, not however without some alarm and a good hearty ducking.

I can only say that if the moor of Rannoch had demonstrated to me the importance of good roads, I was equally convinced of the utility of bridges at Fassifern.

{f. 49 verso}

No opportunity afforded of taking the latitude of Fort William, the weather being continually rainy.

{f. 50}

Tuesday July 1st—Shore of Loch Eil

Mr. Cameron Fassifern was for some time factor on the estate of Locheil. He showed me a plan of part of that estate viz. between Loch Eil and Loch Arkaig, which had been copied from the surveys of the estate made by order of Government during the forfeiture. The boundaries of the several farms are distinctly marked and Mr. Cameron informed me which of them were included in the Shire of Argyle in the upper part of the Glen of Loy. There are several shealings attached to the lower farms and which it became necessary to consider as part of that county in which the respective principal farm was included. The boundary line indeed when drawn has a singular appearance but was sufficiently well ascertained except in the case of the farm of Barr on the Loy. Of this it was only known that one sixth was in the county of Inverness. The exact line of discrimination was unknown. It appeared however, in the opinion of Mr. Cameron and of the ground bailiff that the burn of Alt Leachtan must be the boundary of the Shires, for this leaves much about the sixth of the arable land of that farm in the side of Inverness. And I think this opinion is strengthened by a document I have lately found in the Sheriff clerk's office as shall be mentioned hereafter.

I proceeded with Alex Cameron the ground bailiff on this estate to the head of Loch Sheil. He pointed out to me the march of this estate with Glenfinnan in Inverness and of the latter with Ardgour etc.

It may be observed that a small isle in the head of Loch Sheil is in Argyle Shire.

After the necessary observations I returned to Fassifern.

{f. 51}

Scotch Boundary's [sic]

World	on roller	1–16–0
North America	do	1–10–0
Aus [?]	do	1– 5–0
		4–11–0

{f. 52}

Inverness Academy Oct 12th 1806.

Sir,

I have now completed the delineation of the Boundary of our northern Counties, so far at least as my materials and information have enabled me. I transmit therefore the map last sent down, upon which the Boundary is mark'd. I have coloured the different Shires that the relative situation may be the more readily distinguishable. The slip of the S.W. Sheet I have as desired pasted to the rest. I cannot say that at the particular place where I wanted it, it has given me entire satisfaction. Mr. Arrowsmith has, I fear, paid too great regard to the maps of Argyle and Perth, the surveyors of which (Langlands, Stobie etc) seem to have paid little attention to that District and hardly accessible corner.

I have therefore pasted on a small supplement on which that District is I think tolerably well described, leaving Mr. A to make what use of it he thinks proper. I would only observe that the River Levin is pretty correct while I fear Loch Levin is (on the large map) made rather too long.

The Boundary of the Counties in this neighbourhood, especially Nairn, and Cromarty are so very extraordinary that instead of endeavouring to mark them with the necessary corrections on the small scale of the map, I have thought it more advisable to form a sketch upon double the scale of the whole country from the River Nairn to the Firth of Dornoch. I regret that this has not the neatness which I could wish it to possess, but I hope that it will not be found unuseful. I have delineated upon this the Boundary of Nairn and have added a small supplement for the Estate of Dunmaglass (referred to in the Act for the Inv^{s}. Assessment).

I have also included the different annexations that have been made to Cromarty in that quarter, all of which are I believe derived from very authentic sources with the exception of the Estate of New Tarbat, and part of that of Cadbole. Some other annexations are marked on the large map. I have col^{d} them yellow and orange.

On this sketch I have also exhibited part of the Boundary near Beauly, of the Counties of Ross and Inverness in which there is some little intricacy.

The whole peninsula called the Black Isle affords a specimen of political geography that I think may justly vie with any

{f. 52 verso}

tract in the German Empire for intricacy, for first we have the old Shire of Cromarty, 2nd three district annexations to Cromarty, 3rd a part of Ross, 4th a part of Nairn, 5th two insulated districts of Ross, viz, Findon and Reindown, 6th a small part of Inverness, 7th a disputed tract between Inverness and Ross, and lastly in the centre an extensive common muir to complete the singularity. The inhabitants of the eastern part have for many generations spoken English alone, while those to the west make use of the Galic [*sic*].

There seems a considerable error in the position of the coast of Easter Ross. I regret much that from being confin'd by my professional engagements here, I have not been able to correct this. I have marked an outline that appears to be pretty near the truth and I hope Mr. A will find it not difficult to alter it. In some other places he will see I have departed from the general map as at Beauly, Calder, Dunmaglass and in the Black Isle. This was not done but upon good grounds although 'twould be tedious to insert my Authorities at present.

I have not inserted so many names as might have been done. The truth is I had seldom sufficient information, my time being scarcely sufficient for the chief object. Perhaps Mr. A in some instances is about to insert some that we would think of little value, but indeed this is hardly possible. I would only observe that the system of sheep farming has diminished the number of inhabited places in the Highland Country, so that the more trivial distinctions of neighbouring farms or hamlets, having otherwise the same name, may in many cases perhaps be very well omitted.

I have sketched the outline of a road from Dingwall to the County of Sutherland by a nearer route than has been yet proposed. This is, even at present, pretty much employed, especially as a cattle road, and from what I have been able to learn, it would neither be so difficult in the construction, not liable to such pulls as may be conceived from the mountainous nature of the country through which it passes.

The saving of distance is obvious as we would have nearly a straight line from Beauly to the Bonar.

{f. 53}

I made out a sketch of the District of Strathspey. A considerable part of Inverness is situated there, although insulated from the rest of that County, yet they approach within a quarter of a mile of each other. In like manner a very considerable tract of Morayshire is situated in the upper part of Strathspey some miles distant from the integral part of that County. One farm should be divided between the Counties, but being originally cultivated in alternate ridges 'tis now impossible to discover the boundary.

In some other places the boundary is only known with certainty in the arable lands, the upper country being pastured in common, and as the whole belonged to one proprietor, the limits of Counties have never been defined.

I have endeavoured to exhibit the aspect of the mountains on these sketches. They are the most important objects of Highland geography and serve oftener as the limits of the territory than rivers do.

Upon this sketch I inserted the new military road from Aviemore in Strathspey by Dulsie Bridge to Fort George, and an important alteration upon the military road from the same point to Inverness. Another near Loch Inch on the Spey I have mark'd on the large map. It does not seem to be known to Mr. A. that the road by Drumuachtar and Blair in Athol is the great road to the South.

I have inserted on the large map the proposed outline of a road that is much wished for, to lead from Glen Garry by Glen Spean, Loch Traig, Rannoch, etc to Killin to join the military road to Stirling. This would become the great cattle road and would be the surest means of opening a communication with the Western Highlands.[42]

Another road which seems a branch of this has been proposed to lead from Rannoch to Kings House thereby giving the upper parts of Perthshire a ready communication with the sea at Loch Levin.

Good roads already exist from Rannoch etc. to Perth and from Dunkeld to Montrose. The proposed line would complete this

{f. 53 verso}

and open a direct communication across the Island through the central Highlands. There are only 18 miles to be formed through an open and (though high) a level valley from Rannoch to King's House.

But I am wandering from my proper subject.[43]

In endeavouring to describe the boundaries which I have traversed, I have been led to give a short account of the erection of the several Shires in the North and have procured through the medium of Mr. Hope[44] an extract of some of the Acts of Erection (for we do not, like you, go back to the days of Alfred for that purpose 200 years ago, altho' 4 Northern Counties make part of the Sheriffdom of Inverness and not long before that this County comprehended all that was North of the Grampians). These Acts are sufficiently vague but enable us to discover the causes of the great singularity of some of our outlines. I have not quite finished this but hope to be able to transmit it for you in two or three days.

Meantime I have taken the liberty of writing on the skeleton map a few remarks for Mr. Arrowsmith, referring chiefly to some corrections that are necessary here and there, not indeed of very great importance, yet which I did not wish to omit. Some other little things which it may be necessary to erase I have marked with +++++ inserting near them the necessary alteration.

I regret indeed very much that it was not in my power to proceed to Caithness, especially to examine the ground in dispute at the Ord, but I hope I am not altogether without excuse. I wrote twice to the Convener of the County of Sutherland and received no answer. I called in Dornoch on Mr. McCulloch,

Sheriff Substitute, who informed me that he was in fact Convener and after I had explained my object to him, he said nothing had been done in consequence of the minute transmitted by the Commissioners to Sutherland Shire. He gave me however, a pretty accurate account of the disputed ground at the Ord, a proof having been laid before him of the Sutherland claims. In fine he referred me to Col. Campbell of Combie, factor for the marchioness of Stafford at Dunrobin, whither accordingly I proceeded.

Colonel Campbell, although himself Convener of that County, knew nothing of the object of my mission. He had not been

{f. 54}

in the Country at the meeting 30th April. The minutes had not been forwarded to him. Being himself a stranger in that country, he could give me no information, but proposed we should enquire next day at some individuals who were to be at the Communion in the neighbouring parish Church of Golspie.

The information I procured here from those who seemed best informed served only to convince me of the accuracy of what I already conceived to be the boundary of these Counties. They seemed to think I could procure no further information even by going forward to the Ord.

I had already exceeded the term of absence allowed from my occupations here. I was given to understand that my return was much wished for. I had to investigate the boundary of the County of Cromarty in my way.

I had in every case since leaving Nairn and Strathspey, literally hunted [*sic*] for the necessary information.

I did not incline to pursue the same system to an unlimited extent, begging my way as it were through a country where I now felt myself totally unknown. Was I justifiable therefore in turning my face toward home.

Pardon me for trespassing so long upon your time with a letter of this kind, while,

I have the Honour to be

Sir

Your most obedient Servant

(Signed) Alexander Nimmo.

P.S. I may add that there is a survey of the ground in dispute at the Ord, taken by Provost Brown. The whole extent is hardly a square mile. I have delineated it on the map as well as the small scale admits.

John Rickman Esq.

{f. 55}

(SEE P. XII–XV)

Sketch map 2, four miles to one inch. Captioned by Nimmo: The Heads of the Beaulie River etc. with the county boundary in Glasletir.

There is no Loch Glasletir. Glasletir is a district comprehending the heads of Cannich and, being formerly in the possession of one of the family of Mackenzie, forms part of Ross shire. It extends halfway down Loch Mollardich erroneously named Loch Mayley, and by Ainslie Loch Maddy. The lake marked Loch Glasletir is called L. Lungard from a neighbouring farm. Again, Loch Benevian has been improperly marked Benerack. Observe the Awin Mishkeg on the south of Loch Monar. Observe also Cam Baan in the head of Glen Grivie, also Finneglen these being proposed routes for the great line of communication through this country.

Lochan Uain or the green pool, is represented too large. It is merely a pool of 500 yards by 200 in the bosom of the mountain Mam Soul from which two large glaciers of snow hang into it and keep it always at the freezing temperature.

{f. 56}

The boundary of Ross leaves Loch Monar by a small stream on the south named Alt na Crilie, passes on (over a track to Kintail) to the western shoulder of the mountain Sgur na Lappich the southern brow of which is named Sgur na Clachan Ceal (or the hill of fair stones, specimens of rock crystal being found there). From this it passes eastward by the summits of the mountain and descends the burn called Alt Arderg into the Loch Molardich. Nearly halfway down [it] crosses this lake and to the opposite summits, whence it passes westward by the top of Mam Soul (the highest in that quarter and which by the barometer I find 4020 above the sea, within at most 100 feet of the truth)[45] and makes in fine a complete turn round the head of Glen Grivie as before mentioned.

NB. There is a Sgur na Lappich on the S shoulder of Mam Soul besides the above.

PS. August 14th

I have just seen the copy of the map brought by Mr. Telford.[46] It is beautiful. I find the orthography of a few names in this neighbourhood incorrect and will take the first opportunity of transmitting them corrected. A.N.

Notes to the journal

[1] In July 1803 two separate acts of parliament set up two commissions relating to the development of the Highlands. One was the Parliamentary Commission for Making Roads and Building Bridges in the Highlands of Scotland, the other the Commission for the Caledonian Canal. Thomas Telford was made engineer to both commissions and Charles Abbott was made chairman. Abbott had been chief secretary of Ireland from May 1801 to February 1802.
[2] Part of the cost of the proposed roads and bridges was to be levied on the relevant counties and the beneficial proprietors. This necessitated an accurate survey and map in order to apportion the burden equitably. Nimmo was engaged to carry out the survey of the boundary of Inverness. The Commissioners contracted Aaron Arrowsmith, a London cartographer, to engrave the map, which was published at a scale of 4 inches to the mile in 1807.
[3] Alexander Nimmo, 'Historical statement of the erection and boundaries of the shires of Inverness, Ross, Cromarty, Sutherland and Caithness', in 'Third Report of the Commissioners for the Highland Roads and Bridges', Appendix U, British Parliamentary Papers (BPP) 1807 (100) Vol. III, microfiche (mf) 8.12–3.
[4] John Rickman (1771–1840), 'Lamb's friend the census taker', was a statistician who devised the methods to be employed in the national census and who prepared its various reports between 1801 and 1831. He was private secretary to Charles Abbott and at the same time that the latter was made chairman of both commissions, Rickman was made their secretary.
[5] Sir George Steuart Mackenzie FRS, FRSE (1780–1848) of Coul was an explorer, mineralogist and early geologist. He became a friend of Nimmo and was his proposer when Nimmo was elected FRSE on 1 January 1811. Both men were also members of the Geological Society of London.
[6] Simon Frazer of Farraline (1754–1810), advocate, was Sheriff of Inverness at that time.
[7] This was possibly Archibald Campbell Frazer.
[8] Peter May (1749–1783) was a famous Scottish land surveyor. For further information on him and his work see I.H. Adams (ed.), *Papers on Peter May, land surveyor, 1749–1793* (Edinburgh, 1979).

[9] The true height is 3773 feet (1150m).
[10] Later on (f. 51) Nimmo names the southern face as Sgur na Clacan Ceal (the hill of fair stones).
[11] Robert Dundas, 1771–1851, Keeper of the Signet and MP for Midlothian from 1801. Son of Henry Dundas, he would serve briefly as chief secretary for Ireland in 1809 and become 2nd Viscount Melville in 1811. He was one of the commissioners for the Highland Roads and Bridges.
[12] 'Writer' is an old Scottish term for a lawyer or solicitor (advocate).
[13] Possibly an extract of Ainslie's map *Scotland, drawn from a series of angles and astronomical observations* (Edinburgh, 1789). John Ainslie (1745–1828) was a well-respected cartographer of the time. He does not appear to have published a separate map of Moray.
[14] The provost of Inverness was a member of the Board of Directors of Inverness Academy and therefore Nimmo's nominal employer as rector. He would have known Nimmo well.
[15] A partial eclipse of the sun was anticipated on 16 June 1806.
[16] Commissioners of Longitude, *The nautical almanack and astronomical ephemeris for the year 1806* (London, 1803).
[17] Andrew Mackay, *The theory and practice of finding the longitude at sea or land: to which are appended various methods of determining the latitude of a place and the variation of the compass; with new tables* (2nd edn, Aberdeen, 1801). The first edition appeared in 1793. The third edition did not appear until 1809, so Nimmo must have used either the first or second edition.
[18] General George Wade, 1673–1748, was born of Cromwellian stock in Westmeath, Ireland and joined the British Army in 1690 where he had a distinguished career at home and abroad. Between 1725 and 1737 he directed the construction of about 250 miles of military road through the Highlands, including up to 40 bridges. He rose to be commander-in-chief of the entire British army in 1744.
[19] By the compass.
[20] See note 3 above.
[21] The erasure stated 'we also proceeded'.
[22] Braeriach at 4252 feet (1296m) is indeed higher than Cairn Gorm at 4085 feet (1254m).

[23] 'Stand fast, Craig Elachie!' was the battle-cry.
[24] Richard Kirwan (1733–1812), chemist and mineralogist, was born in Co. Galway, Ireland. He became president of the Royal Irish Academy in 1799. In 1787 he published a treatise entitled *An estimate of the temperature of different latitudes.*
[25] Hector Boethius (*c.*1465–1536), *Scotorum historiae* (First published 1527). Published in English as *The description of Scotlande written by H. Boethius and translated into the Scottish speech by John Bellendon and now finally into English by W.H [William Harrison] 1577.*
[26] L. Shaw, *The history of the province of Moray* (Edinburgh, 1775).
[27] Robert Sibbald, *The history, ancient and modern, of the sheriffdoms of Fife and Kinross: with the description of both and of the Firths of Forth and Tay* (Cupar Fife, 1803).
[28] According to Robin Smith in *The making of Scotland: a comprehensive guide to the growth of its cities, towns and villages* (Edinburgh, 2001), Kingussie was laid out by the Duke of Gordon in 1799, had a post office by 1802 and a woollen mill by 1805. It grew and developed after the road through it from Perth to Inverness was opened and especially when the railway came. Today it is mainly a tourist village.
[29] Now called Beinn Udlamain, this mountain is 3316 feet (1011m) high.
[30] Nimmo variously refers to Loch Ericht as Erucht, Eroch, Errich.
[31] Dal n' Lynchart appears to have been where Benalder Lodge is today.
[32] James Stobie, *Map of Perthshire and Clackmanann* (London, 1783).
[33] That is, 480 feet. Today, Loch Ericht, which is now dammed at both ends, has a maximum depth of 512 feet (156m).
[34] Nimmo described, in a brief comment, part of the experiments he carried out with Simon Fraser of Foyers in 1804 on Loch Ness in his article entitled 'On the application of the science of geology to the purposes of practical navigation', published in *Transactions of the Royal Irish Academy* XIV (1825), 39–50.
[35] Ben Aaler, now called Ben Alder, is 3766 feet (1148m) high; Aonach Mor is 4006 feet (1221m); Beinn Udlamain is 3316 feet (1011m). Nimmo's heights are given in yards above the level of the loch, which is located 1153 feet (351m) above sea level. By adding this to Nimmo's values, his estimated heights are within about 2

per cent of the correct values for Ben Alder and Beinn Udlamain, but he underestimated Aonach Mor by about 25 per cent.

[36] This probably refers to the Caledonian Canal, then in course of construction through the Great Glen.

[37] He probably meant here Ben Cruachan, 3694 feet (1126m) high on the banks of Loch Awe.

[38] There is also a sea loch in Mull named Loch an Scridain.

[39] George Langlands and Sons, *Map of Argyleshire* (Campbeltown, 1801).

[40] General William Roy F.R.S. (1726–90), military surveyor and engineer, carried out the Military Survey of Scotland between 1747 and 1755 while still a young man. The result was published as William Roy, *The great map: the Military Survey of Scotland 1747–55* (Edinburgh, 1793) and is still regarded as a useful reference document, a digitised copy of which can be consulted through the National Library of Scotland.

[41] John Telford was resident engineer (under Thomas Telford, no relation) at the western end of the Caledonian Canal. He died the year after Nimmo's survey. The other person, Mr Donkim, has not been positively identified and is not named in A.W. Skempton, *Biographical dictionary of civil engineers in Great Britain and Ireland, Volume 1: 1500–1830* (London, 2002).

[42] In 1810 Telford proposed this road both in the 'Fifth Report of the Commissioners for the Highland Roads and Bridges' (London, 1810/11) and in a separately published booklet, 'Report and estimates relative to a proposed road in Scotland from Kyle Rhea in Inverness-shire to Killin in Perthshire by Rannoch Moor' (London, 1810). In these, Telford credited Nimmo only with some calculations as to the likely benefit of the road. However, it is not clear whether or not Telford's proposal was based on Nimmo's proposal here, or whether Telford had first made the proposal before 1806, but not published it. On the face of it, Nimmo's reference here to this road makes the proposal his own and not Telford's.

[43] Nimmo knew full well that his employers on the survey were the Commissioners for the Highland Roads and Bridges and Rickman was their secretary. So his 'wandering from my proper subject' was not at all inappropriate or ill-conceived: it resulted in his being con-

tracted subsequently to survey the route from High Bridge to Killin for the Commissioners in 1809, although no such road was ever built.

[44] James Hope, an Edinburgh lawyer, Writer to the Signet, became the Agent of the Commission for Highland Roads and Bridges, dealing with its legal and financial business in Scotland. As Haldane wrote, 'Hope in Edinburgh wrestled with difficulties calling for skill in law and finance and no less for understanding of human nature, while in London Rickman held in his hands the threads of the whole complex undertaking, keeping them from enravelment and confusion', A.R.B. Haldane, *New ways through the Glens: Highland road, bridge and canal makers of the early nineteenth century* (Edinburgh, 1962).

[45] The true height of Mam Soul or Maam Suil, also known as Mam Sodhail, is 3875 feet (1181m). Nimmo's estimate is within 145 feet of this i.e. within 4 per cent of the true value.

[46] Aaron Arrowsmith, *Map of Scotland, constructed from original materials*, National Archives of Scotland, RHP 14008 (London, 1807).

BIBLIOGRAPHY

Primary Source

Nimmo, Alexander (no date) Letter from Alexander Nimmo to the Inverness Academy directors, with inserted letter from Alexander Campbell, English teacher. Highland Council Archive CI/5/8/9/3/1/1.

Nimmo, Alexander (no date) 'Geological Map of Connemara, Ireland'. Notes captioned by George Bellas Greenough 'Nimmo', 'Connemara'. Manuscript map archived in the Geological Society of London, LDGSL 999. (A coloured reproduction of this map is given in Wilkins, *Alexander Nimmo.*)

Nimmo, Alexander 1805/6 'Thumbnail sketch of Fort George'. Kingussie. Highland Folk Museum Collection, ref. cc17, 1864.

Nimmo, Alexander 1806 'Journal along the North East and South of Inverness-shire. Ends at Fort William'. Manuscript, Inverness. Adv MS 34.4.20, National Library of Scotland. Reproduced in this volume. Dublin. Royal Irish Academy.

Nimmo, Alexander 1807 'Historical statement of the erection and boundaries of the shires of Inverness, Ross, Cromarty, Sutherland and Caithness', in 'Third Report of the Commission for the Highland Roads and Bridges', Appendix U, British Parliamentary Papers (BPP) (100) Vol. III, mf 8.12–3.

Nimmo, Alexander and Thomas Telford 1811 'Bridge' in D. Brewster (ed.), *The Edinburgh encyclopaedia*. Edinburgh. William Blackwood.

Nimmo, Alexander 1813/14 'The Report of Mr. Alexander Nimmo on bogs in the barony of Iveragh in the County of Roscommon', Appendix 5 in 'Fourth Report of the Commissioners for the Bogs of Ireland, BPP (131) Vol. VI, Pt 2, mf 15.33–6. Hereafter 'Fourth Bogs Report'.

Nimmo, Alexander 1813/14 'The Report of Mr. Alexander Nimmo on the bogs in that part of the County of Galway to the west of Lough Corrib', Appendix 12 in 'Fourth Bogs Report'.

Nimmo, Alexander 1813/14 'The Report of Mr. Alexander Nimmo on the remaining bogs in various parts of the Counties of Kerry and Cork', Appendix 6 in 'Fourth Bogs Report'.

Nimmo, Alexander 1825 'Evidence', in 'Minutes of evidence taken before the Select Committee of the House of Lords to examine into the nature and extent of the disturbances which have prevailed in those districts of Ireland which are now subject to the provisions of the Insurrection Act and to report to the House', BPP (200) Vol. VII. mf 27.60–63, 179. Hereafter 'House of Lords Report'.

Nimmo, Alexander 1825 'On the application of the science of geology to the purposes of practical navigation', *Transactions of the Royal Irish Academy* XIV, 39–50.

Nimmo, Alexander 1826 'Evidence', in 'Minutes of evidence before the Committee on the Norwich and Lowestoft Navigation Bill', BPP (396) Vol. IV, mf 28.24–6.

Nimmo, Alexander 1830 'Report on the progress of the public works in the western district of Ireland in the year 1829', in 'Public Works Ireland', BPP (199), Vol. XXVII, mf 32.194, 3–4.

Anonymous 1832 Obituary of Alexander Nimmo, *Galway Advertiser*, Vol. XIV (5), 28 January.

Ainslie, John 1789 *Scotland, drawn and engraved from a series of angles and astronomical observations*. Edinburgh. J & J Ainslie and W. Faden.

Arrowsmith, Aaron 1807 *Map of Scotland, constructed from original materials*. London. Published by Aaron Arrowsmith. National Archives of Scotland, RHP 14008.

Arrowsmith, Aaron 1809 *Memoir relative to the construction of the map of Scotland*. London. Published by Aaron Arrowsmith in 1807. Reprinted in Edinburgh 1809. Edinburgh Geological Society. National Archives of Scotland, GD9/40/1.

Arrowsmith, Aaron 1813 *Map of the hills, rivers, canals, and principal roads, of England and Wales: upon a scale of six miles to an inch. Exhibiting most of the places whose situation has been ascertained by the stations and intersections of the trigonometrical survey*. London. Published by Aaron Arrowsmith.

Boethius, Hector 1527 *Scotorum historiae*, first published in English 1577 as *The description of Scotlande written by H. Boethius and translated into the Scottish speech by John Bellendon and now finally into English by W.H [William Harrison]*.
Brewster, David (ed.) 1811 *The Edinburgh encyclopaedia*. Edinburgh. Blackwood.
Chambers, Robert 1856 *A biographical dictionary of eminent Scotsmen, Volume V*. Glasgow. W & R Chambers.
Commissioners of Longitude 1803 *The nautical almanack and astronomical ephemeris for the year 1806*. London. Printed by T. Bensley and sold by Payne and MacKinlay.
Earl of Morton's Papers 1810 Edinburgh. National Archive of Scotland, GD150/3420.
Inverness Royal Academy 1792 Opening announcement in *Edinburgh Evening Courant*, 21 June. Edinburgh.
Inverness Royal Academy 1808 Visitation Committee Report March. Edinburgh. National Archive of Scotland, GD128/34/1/36.
Inverness Royal Academy 1810 *Inverness Journal* Advertisement 8 June 1810.
Inverness Royal Academy 1835 'Rector's report of the state of the Inverness Academy to the directors, at their Annual Meeting on 30th April 1835,' Inverness Royal Academy Archive, B2.
Kirwan, Richard 1787 *An estimate of the temperature of different latitudes*. London. Printed by J. Davis for P. Elmsly in the Strand.
Knox, John 1787 *A tour through the Highlands of Scotland, and the Hebrede Isles, in 1786*, reprinted 2009. London, Edinburgh, Glasgow. J. Walton Subjects. Scotland Hebride Isles History.

Langlands, George & Sons 1801 *Map of Argyleshire.* Campbelstown, n.s. National Library of Scotland, Maps, shelfmark EMS s. 326.

Loch, J. 1820 *An account of the improvements on the estates of the Marquess of Stafford.* London. Longman et al.

Mackay, Andrew 1801 *The theory and practice of finding the longitude at sea or land: to which are appended various methods of determining the latitude of a place and the variation of the compass; with new tables,* 2nd edn. Aberdeen. Printed for the author by J. Chalmers & Co.

Mitchell, Joseph 1883 *Reminiscences of my life in the Highlands, Volume 1.* Chilworth. 1st edn printed privately for the author at the Gresham Press. 1971 reprinted in two volumes.

Mudie, R. 1842 'Notes by the conductor', in R. Mudie (ed.) *The surveyor, engineer and architect for the year 1842.* London. Bell & Wood.

Playfair, John 1804 in 'Minutes of the Royal Society of Edinburgh', 3 December. Edinburgh. National Library of Scotland, Acc 10000/4.

Rickman, J. (ed.) 1838 *Life of Thomas Telford, civil engineer, written by himself.* London. Payne, Foss, J & G Hansard and Sons.

Roy, William 1785 'An account of the measurement of a base on Hounslow Heath', *Philosophical Transactions of the Royal Society of London* 75, 385–478.

Roy, William 1793 and 2007 *The great map: the Military Survey of Scotland 1747–55.* Edinburgh. Birlinn.

Shaw, L. 1775 *The history of the province of Moray.* Edinburgh. W. Auld.

Sibbald, Robert 1803 *The history, ancient and modern, of the sheriffdoms of Fife and Kinross: with the description of both and of the firths of Forth and Tay*. Cupar Fife. R. Tullis.

Stobie, James 1783 *Map of Perthshire and Clackmannan*. London, s.n. National Library of Scotland, Maps, shelfmark EMS.b.2.30.

Telford, Thomas 1802/3 'A survey and report of the coast and Central Highlands of Scotland, made by command of the Right Honorable the Lords Commissioners of Her Majesty's Treasury, in the autumn of 1802, by Thomas Telford, Civil Engineer, Edinburgh FRS'. BPP (45).

Thomas Telford to Andrew Little 18 February 1803, quoted in A.Gibb *The story of Telford*, 71-2.

Telford, Thomas 1810 'Report and estimates relative to a proposed road in Scotland from Kyle-rhea in Inverness-shire to Killin in Perthshire by Rannoch Moor'. London. Luke Hansard & Sons.

'Fifth Report of the Commissioners for the Highland Roads and Bridges'. Appendix U, BPP (112) Vol. IV, mf 12.24.

Telford, Thomas 1819 'Evidence' in 'Second Report from the Select Committee on the State of Disease and Condition of the Labouring Poor in Ireland'. BPP (314) Vol. VIII, mf 20.67.

Thompson, John 1832 *The atlas of Scotland*, reprinted 2008. National Library of Scotland. Edinburgh. Birlinn.

Wellesley Papers 1823 Letter from H. Goulburn to Lord Wellesley, 23 May. London. British Library, Add 37301 f 90.

Wye Williams, Charles 1832 Obituary of Alexander Nimmo, *Dublin Evening Post*, 28 January (Original manuscript in University College London, Greenough Papers).

Secondary Source

Adams, I.H. (ed.) 1979 *Papers on Peter May, land surveyor, 1749–1793*, Scottish History Society, 4th series, Vol. 15. Edinburgh. Constable.

Anderson, C.J. 2009 'State imperatives: military mapping in Scotland, 1689–1770', *Scottish Geographical Magazine* 125 (1), 4–24.

Ashton, T.S. 1948 *The Industrial Revolution 1760–1830*, revised edn 1962. London. Oxford University Press.

Birse, R.M. 1983 *Engineering at Edinburgh University, a short history 1673–1983*. Edinburgh. University of Edinburgh.

Cregeen, E.R. 1964 *Argyll estate instructions: Mull, Morvern, Tiree (1771–1805)*. Edinburgh. Scottish History Society.

Fenyo, K. 2000 *Contempt, sympathy and romance: Lowland perceptions of the Highlands and the clearances during the famine years, 1845–1855*. East Linton. Tuckwell Press.

Fleet, Christopher 2005 'James Stobie and his surveying of the Perthshire landscape', *History Scotland* 5 (4), 40–47.

Gibb, A. 1935 *The story of Telford*. London. Alexander Maclehose.

Gibson, R. 2007 *The Scottish countryside: its changing face, 1700–2000*. Edinburgh. John Donald, National Archives of Scotland.

Haldane, A.R.B. 1962 *New ways through the Glens: Highland road, bridge and canal makers of the early nineteenth century.* Edinburgh. Thomas Nelson & Sons.

Harley, J.B. 1987 'The map and the development of the history of cartography' in J.B. Harley and D. Woodward (eds), *The history of cartography, Volume 1: Cartography in Prehistoric, Ancient, and Medieval Europe and the Mediterranean,* 1–42. Chicago. University of Chicago Press.

Harley, J.B. 1964 'The Society of Arts and the surveys of English counties, 1759–1809', *Journal of the Royal Society of Arts* 112, 43–6; 119–24; 269–75; 538–43.

Hunter, James 1976 *The making of the crofting community.* Edinburgh. John Donald.

Hunter, James 2011 'The Scottish Highlands and Ireland in the time of Alexander Nimmo' in this volume. Dublin. Royal Irish Academy.

MacKenzie, W.M. 1905 *Hugh Miller: a critical study.* London. Hodder & Stoughton.

Moir, D.G. (ed.) 1973 and 1983 *The early maps of Scotland to 1850, Volume 1.* Edinburgh. Royal Scottish Geographical Society.

Parry, M.L. and Slater, T.R. 1980 *The making of the Scottish countryside.* London. Crook Helm.

Robinson, Mary, President of Ireland 1997, lecture. Skye, Glasgow. Oraid Sabhal Mòr Ostaig Gaelic-medium college.

Ruddock, Ted 2002 'Alexander Nimmo' in *Biographical dictionary of civil engineers in Great Britain and Ireland, Volume 1: 1500–1830.* London. Institution of Civil Engineers.

Skempton, A.W. 2002 *Biographical dictionary of civil engineers in Great Britain and Ireland, Volume 1: 1500–1830.* London. Thomas Telford, Institution of Civil Engineers.

Smart, R.N. 2004 *Biographical register of the University of St Andrews, 1747–1897.* St Andrews. University of St Andrews.

Smith, Robin 2001 *The making of Scotland: a comprehensive guide to the growth of its cities, towns and villages.* Edinburgh. Canongate.

Wilkins, Noël P. 2009 *Alexander Nimmo, master engineer, 1783–1832: public works and civil surveys.* Dublin. Irish Academic Press.

Wills, Virginia (ed.) 1973 *Reports on the Annexed Estates, 1755–1769.* Edinburgh. HMSO.

Withers, C.W.J. 2002 'The social nature of map making in the Scottish Enlightenment, *c.*1682–1832', *Imago Mundi* 52, 46–66.

NOTES ON THE AUTHORS

Christopher Fleet

Christopher Fleet is Senior Map Curator at the National Library of Scotland, Edinburgh, where he has been based since 1994. At the Library, his main responsibilities have been in scanning and delivering historical mapping over the Web, as well as in managing digital map and data applications. His research interests are focused on the Library's cartographic collections, particularly relating to Scotland, from the earliest mapping by Timothy Pont in the late sixteenth century, through to modern digital mapping today. He has published widely on these subjects.

James Hunter

Professor James Hunter, CBE FRSE, is Director of the Dornoch-based UHI Centre for History, a key component of the prospective University of the Highlands and Islands. The author of 11 books on the Highlands and on the region's worldwide diaspora, he has also been active in the public life of the area. In the mid-1980s he became the first director of the Scottish Crofters Union, now the Scottish Crofting Federation. More recently, he served for six years as chairman of Highlands and Islands Enterprise, the north of Scotland's government-funded development agency. In the course of a varied career, Professor Hunter has also been an award-winning journalist and broadcaster.

Robert Preece

Robert Preece was born in Edinburgh and educated at George Heriot's School and the University of Edinburgh. After completing teacher training, he took a job as teacher of Geography at Inverness Royal Academy, where he soon became Principal Teacher of Geography. He also more recently taught media studies. Since retirement, his many interests include acting as Honorary Archivist to the school, and he is currently completing the first history ever to be published on the school. This has involved major research throughout Britain, although mainly in Scotland. He has a life-long involvement, first as a youngster and then as a leader and commissioner, with the Scout Association. As a volunteer he also looks after the local Scout archive, and parts of the archive of the Diocese of Moray, Ross and Caithness in the Scottish Episcopal Church.

Noël P. Wilkins

Professor Noël P. Wilkins is Emeritus Professor of Zoology at the National University of Ireland, Galway.

He worked for nine years in the Marine Laboratory of the Department of Agriculture and Fisheries for Scotland in Aberdeen. Transferring to Galway in 1970 he lectured and researched in Zoology, concentrating on the physiology and genetics of marine organisms. He has published over 80 research papers on these topics. His later interests were in the history of fisheries, in which he first encountered the work of Alexander Nimmo on Irish piers and harbours. He is the author of four books on fisheries and his latest book, published in 2009, is a study of the life and works of Nimmo, entitled *Alexander Nimmo, master engineer, 1783–1832: public works and civil surveys.*

Topographical and Placename Index

Numbers refer to folios in the original manuscript V=verso

RIVERS

MOUNTAINS

DISTRICTS

CITIES, TOWNS, VILLAGES

PARISHES

FARMS

OTHER